Smart Nanoparticles Technology

Smart Nanoparticles Technology

Editor

Hua Fen Han

scitus academics

Smart Nanoparticles Technology
Edited by **Hua Fen Han**

ISBN: 978-1-68117-670-3
Library of Congress Control Number: 2016946536

SCITUS Academics LLC,
616, Corporate Way, Suite 2, 4766,
Valley Cottage, NY 10989

www.scitusacademics.com

Notice

Reasonable efforts have been made to publish reliable data and views articulated in the chapters are those of the individual contributors, and not necessarily those of the editors or publishers. Editors or publishers are not responsible for the accuracy of the information in the published chapters or consequences of their use. The publisher believes no responsibility for any damage or grievance to the persons or property arising out of the use of any materials, instructions, methods or thoughts in the book. The editors and the publisher have attempted to trace the copyright holders of all material reproduced in this publication and apologize to copyright holders if permission has not been obtained. If any copyright holder has not been acknowledged, please write to us so we may rectify.

Printed in United States of America on Acid Free Paper

Preface

Because of their versatile application, environmentally responsive microgels have been subjects of great interest in the past two decades. Such microgels are often referred to as "intelligent" because their properties allow them to respond to external stimuli in a certain way. Some "intelligent" materials can therefore be adapted to different parameters such as temperature and ion strength, light, pH and magnetic fields. These methods are applied in various fields such as drug delivery, catalysis, biomaterials and chemical separation and biosensing. The application and synthesis of mini-emulsion inorganic and polymer nanoparticles is increasingly popular in industry and academia. The formulation of polymer latexes in environmentally friendly water without the use of organic solvents is generally of great importance for safety, environmental and health reasons. One formulation of nanoparticles that can be used in material science or possibly in biomedical applications is permitted by the mini-emulsion process. Nanoparticles and their applications have dramatically diverted science into a new philosophy in recent years. When nanoparticles were formed, the quality of many conventional materials changed. Nanoparticles have a higher surface area per weight than larger particles, which leads to reactivity and success. Smart Nanoparticles Technology explains comprehensive aspects of nanoparticles, showing their advantages and advanced development in many applications. This book describes the inherent properties and applications of nanoparticles in different fields. It comprises a number of well-researched studies that help students and nanotechnology experts.

Table of Contents

Chapter 1

Polymer Nanoparticles for Smart Drug Delivery

Devasier Bennet[1] and Sanghyo Kim[1, 2]

[1] Department of Bionanotechnology, Gachon University, Bokjeong-Dong, Sujeong-Gu, Seongnam-Si, Republic of Korea

[2] Graduate Gachon Medical Research Institute, Gil Medical Center, Inchon, Republic of Korea

1. INTRODUCTION

In the recent decades, polymers are widely used as biomaterials due to their favorable properties such as good biocompatibility, easy design and preparation, a variety of structures and interesting bio-mimetic character. Especially in the field of smart drug delivery, polymer played a significant role because it can deliver therapeutic agents directly into the intended site of action, with superior efficacy. The ideal requirements for designing nano-particulate delivery system are to effectively be controlled particle size, surface character; enhance permeation, flexibility, solubility and release of therapeutically active agents in order to attain the target and specific activity at a predetermined rate and time. The smart drug delivery systems have been successfully made by the advances in polymer science in the bio-nanotechnology field. Recently, these advances have been found in various medical applications for nano-scale structures in smart drug delivery. The smart drug delivery systems should possess some important feature such as pre-scheduled rate, self controlled, targeted, pre-determined time and monitor the delivery. The smart drug delivery system enhances the polymer nanoparticle better stage to their therapy regimen. They are drug carriers of natural, semi-synthetic, and synthetic polymeric nature at the nano-scale to micro-scale range. The polymeric particles are collectively named as spheres and capsules. The most of the polymeric nanoparticles with surfactants offer stability of various forms of active drugs and have useful to smart release properties. There are numerous biological applications have been reported for the nano-scale to micro-scale sized particles, such as site-targeted, controlled, and enhanced bioavailability of hydrophobic drugs [1-4]. Due to the nanoparticles size the drugs have been targeting into various applications, such as, various cancers targeting has been shown to be promising [5]. Moreover, polymeric particles proved their effectiveness in stabilizing and protecting the drug molecules such as proteins, peptides, or DNA molecules from various environmental hazards degradation [2-4, 6, 7]. So these polymers are affording the potential for various protein and gene delivery. Numerous methods had been available to fabricate nanoparticles; it depends on the physical and chemical properties of polymer and active ingredients. Most of the formulation techniques involve different mechanisms such as using organic solvents, temperature, ultra-sonication and mechanical agitation which can degrade the pharmaceutical active ingredients. So the nano-particulate system can be developed to consider the formulation methodology should not damage the active pharmaceutical ingredients. There are numerous biodegradable and biocompatible polymers with different physicochemical characters are offered to prepare smart nanoparticles, those polymeric nano-carriers can be natural or semi-synthetic or synthetic. Those nanoparticles can enhance the systemic circulation half-life and minimize unwanted internalization and prevents the denaturation of the therapeutically active moiety and could use to deliver the target agents. Several polymer systems are approved by the U.S. Food and Drug Administration (FDA) for human use. It is the belief that when inventions in fabrication can catch up with those in materials, design and development of drug delivery system can enter a new generation of enhancing clinical healthcare.

The most recent advances in the uses of carriers for sustained and targeted delivery, micro and nano fabricated self-regulated devices [8], bio-recognizable systems; micro-needles for transdermal drug delivery have shown the flexibility and enhanced permeability of these polymeric materials. Ultimately the goal in smart drug delivery is the emergence of a micro and nano-fabricated therapeutic drug release device with the capacity to enough hold and release of various active agents on demand. In modern system the micro-electro-mechanical systems give a distinctive possibility to produce micro-fabricated biomedical devices for different intentions, from implantable systems to lab-on-a-chip systems. The constant and prolonged drug release micro-fabricated systems have the several benefits, such as many active ingredients could be stored in an nano form within the system and sustainably released, the drug release is initiated by the dissolution and disintegration of outer membrane barrier by an mechanical/electric stimuli, the most potential drugs could be released more specifically with this technique, the complex drug release system such as simultaneous stable and periodically could be attained for local therapy by the micro-fabricated system; it can be achieved in high or low dose of drugs at the targeted site and increase the stability of drugs by the membrane barrier for preventing water diffusion into the reservoirs [9]. Owing to the advanced scientific sophistication of the controlled drug release system that has been achieved till now, or that are in dynamic progress, this delivery model can be categorized into various classes. The controlled drug delivery systems can be categorize four main mode of drug delivery, such as (1) rate-programmed drug delivery, where drug diffusion from the system has follow a specific release rate profile, (ii) activation-modulated drug delivery, where the drug release is induced by various factors such as physical, chemical electrical or biochemical modules, (iii) feedback-regulated drug delivery, where the rate of release is determined by biochemical substance (triggering agent) concentrations, it is dependent on the concentration exhibit in the target and (iv) site-targeting drug delivery systems, this is a complex process that consists of multiple steps of diffusion rate and partitioning for the rate of drug release is regulated by the specific targeting moiety, solubilizer and drug moiety. This chapter will brief discussion on recent innovative nano-fabrication methods for novel drug delivery system. Also, highlights some of these new technologies and consider their possibility ongoing clinical transformation of nanoparticles, which the particles are well-controlled formulated. This chapter will be followed by a more detailed novel drug delivery system development from a polymeric material viewpoint and their various bio-applications will be covered without attempting to all the work that has been done in this field.

Over a decade, investigators have appreciated the enrichment of potential uses of bio-nanotechnology in offering huge advancements in novel drug delivery and targeting. The novel drug delivery platform that provides diminishes toxicity and enhances therapeutic efficacy gives most possible benefits to clinical levels. In approaches to drug delivery systems the route of administration is one of the crucial roles of drug targeting. These nanoparticles can be used for various routes, including oral, nasal, transdermal, parenterals, pulmonary, ocular, etc. Nonetheless, the oral route is most convenient, preferred, and in several cases, also its cost-effective, but it does not cross easily some biological barrier; also easily degraded by various body fluids, then rapid hepatic clearance and other organs. So the drug delivery systems focus on overcoming the various membrane barriers, such as the blood brain barrier, tight junction barrier, to achieve the effective drug target and enhance the efficacy. To find an alternative and satisfiable route of administration for the effective drug delivery system should overcome the digestive tract problems, where the degradation could take place via acid-hydrolysis, enzymatic degradation and bacterial fermentation in the alimentary canal. This chapter will cover the more detailed novel route of administration and development from a polymeric material viewpoint and their brief discussion will be covered without attempting to all the work that has been done in this field.

2. GENERAL METHODS FOR POLYMERIC NANOPARTICLES PREPARATION

Recently, various kinds of polymers are used to prepare the polymeric nanoparticles, among this all polymer biodegradable polymers and their co-polymers such as di-block, tri-block, multi-block or radial block copolymer structures have been generally used to prepare polymeric nanoparticles and to encapsulate the active ingredients. These multi-functionalized polymeric nano-carriers include micelles, capsules, platelets, fibers, spheroids colloids, dendrimers, core-shells, nanoparticle incorporated polymer matrixes, etc. The first polymeric nanoparticles were developed between the year of 1960 to 1970 for the therapeutic application, and this were Micelles[10-12].The micelles are formed by polymerisation methods, commonly the formation of polymer nano-carriers during the polymerization of monomers [13-16]. Then the various advanced polymerization techniques have been developed for the preparation polymeric based nanoparticles, and the nanoparticles were stabilised using various surfactants [1, 9]. The stabilised drug loaded nanoparticles consist of drug and non-toxic biocompatible polymer with stabilizing agents, the biocompatible polymer is either biodegradable or non-biodegradable. Numerous techniques are available for the preparation of the polymeric nanoparticles and mainly top-down and bottom up processes. The polymer nanoparticle drug carriers can be further categorized into nano/micro-capsules and nano/micro-spheres depends on the size and structure [1, 9, 17-19]. The fine particles are 100 - 2,500 nm and ultrafine particles are 1 to100 nm in size, and are collectively known as nanoparticles. 50 to 300 nm sized nanoparticle have been prepared by emulsion polymerization method [20]. Drawbacks in polymerization techniques are evolving noxious factors such as toxic, reactive residues, un-reacted monomers, the risk of a chemical reaction and the formation of unwanted oligomers [1], and these drawbacks are overcome by using preformed polymers for the polymerization process [1]. Generally the drug loaded nanoparticles were prepared by dissolving the drug and polymer into the water-immiscible organic solvents and producing a nano-emulsion, as an example by probe-sonication method. The organic solvent is removed by using elevated temperature or reduced pressure [21-23], as an example of rotary evaporation method, and the nanoparticle is washed and collected by certification. Followed by various changes and improvements of the emulsification techniques have been reported [24-29]. For example, the sonication process is a crucial step in the preparation of the sensitive drug loaded nanoemulsion, and the sonication process can increase the temperature, that leads to inactivate the active ingredients. In order to avoid the problems researchers utilized an on/off cycle to maintain a low temperature. Other examples of general methods to prepare the drug polymer nanoparticle are described in the Figure 1. The biodegradable polymeric nanoparticles are commonly prepared by five different techniques such as emulsification-solvent evaporation, solvent displacement, salting-out, emulsification-solvent diffusion and double emulsion solvent evaporation. The synthesizing methods include salting-out method [1, 30, 31]; it is based on the separation of a water miscible solvent from aqueous solution through the salting out effect, solvent displacement method [1, 32-34], phase separation method [35], evaporation precipitation [36, 37], antisolvent precipitation and electrospray methods [38].

Also, many approaches have been developed for the drug particle size reduction (increase in the surface) to the nanometer size range. For size-reduction, high pressure homogenization or wet bead milling is frequently used technique to produce reduced size nanoparticle [39-43]. Among these the high-pressure homogenization has been shown to be effective methods to produce size reduction particle. Moreover, its need sophisticated equipment to resist increasing pressures and temperature. Then, in order to obtain dried polymeric nanoparticle formulations researchers used various drying techniques such as atmospheric freeze drying, spray freeze drying, vacuum freeze drying, and lyophilisation. The uniformity of spray-dried nanoparticle is better than a freeze-dried nanoparticle. Moreover the lyophilisation and spray-drying are used to prepare the nanoparticle [44, 45], these nanoparticles easily tends to aggregates. Also the polymeric nanoparticles have also been synthesized by supercritical fluid techniques [46-52]. This method can get a dry product without any solution, also no need additional drying stages, but the supercritical fluid can swell some of the polymers and act as a softener, extender, and lubricant, which lead to aggregation. Moreover, this method is not easy to get the mono-dispersed multi-component particles because of different kinetics [52]. Nanoparticles prepared by spray-drying technique are one-step based on the conversion of a droplet to a dry particle by evaporation [53-55]. These one-step techniques have been revealed that the nanoparticle could be prepared without any problems [56-58], and the drug content in the particles is almost high [59], but produce an amorphous residual structure. In all above technique induce some unwanted noxious factors, as well as the organic solvents used in the preparations are increasing the risk of pharmaceutical application, also the increased processing time leads to microbial contamination [60,61, 62]. Understanding the all risk factors, recently the modern instrument provides a promising and viable platform for the preparation polymeric nanoparticles.

3. MODERN METHODS FOR PREPARATION OF POLYMERIC NANOPARTICLES

Recently, the polymeric nanoparticles have emerged as a most promising and viable technology platform for recognizing the targeted, environment-responsive and, multi-functional with navigated controlled drug delivery system. Polymer in smart drug delivery is a rapid-emerging new technological discipline in which various therapeutic applications of nano products are expected to overcome the patient complaints in healthcare. Smart delivery will give new solutions for therapeutic interventions. There is great interest from the beginning in smart medicine of advanced and well-characterized bionanotechnological products that will be especially effective in fighting diseases like cardiovascular diseases [63], diabetes [64], cancer [65, 66], aging [67, 68], some chronic metabolic syndrome and various degenerative diseases and disorders [69, 70]. For example, the innovative smart polymers with nano-particulate drug-delivery systems can obviously advances in therapeutics by guiding the drugs to target cells and reducing the adverse-effect/side-effect on well being. At present, some of the smart polymer with multi-functioned nanoparticle system approaches in clinical trials, and it shows promising outcome. Certainly the morbidity and mortality rate of disease affected patients could improve their lifestyle by the early course of smart therapeutic intervention. This smart intervention can be attained by developing high sensitivity and reliable smart drug delivery.

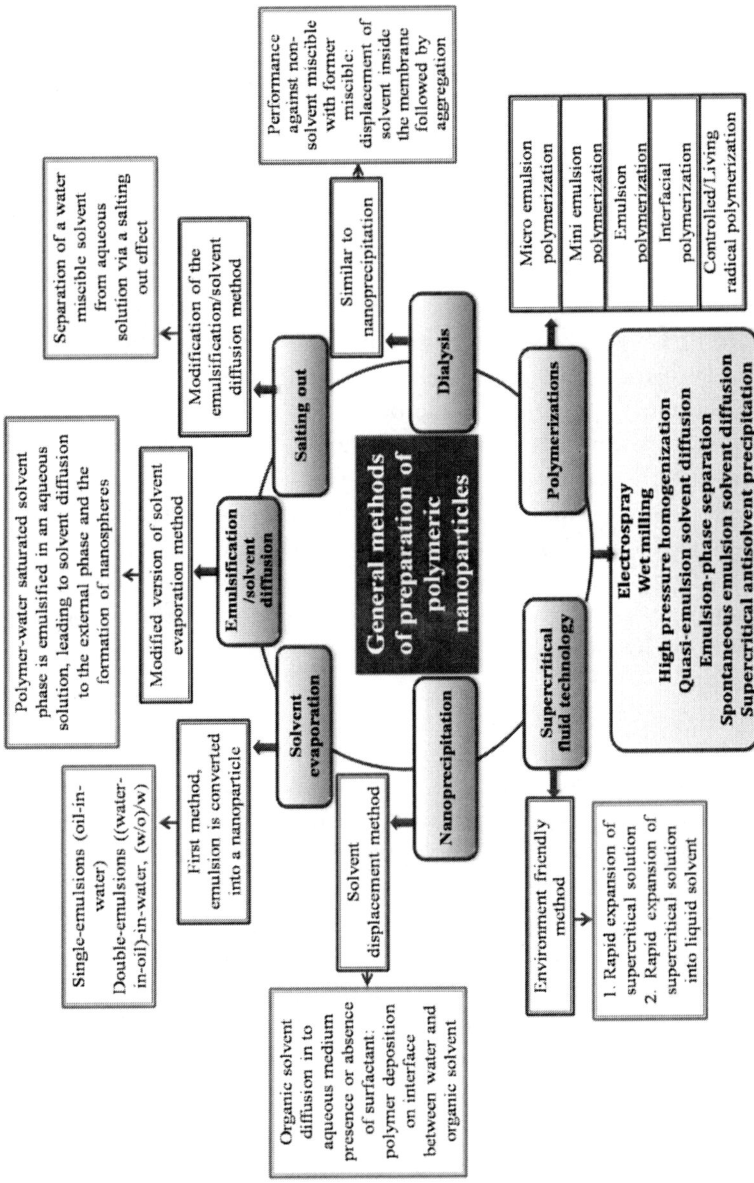

Figure 1. General methods of preparation of polymeric nanoparticles and their principle involved in the mechanisms

The rapid advancement in the above direction has been made with the initiation and development of more advanced alternative nanofabrication techniques to produce structures in various nano-scales level of controlled manners. Drug loaded polymeric nano-systems can provide controlled release of both hydrophilic and hydrophobic drugs over a long period of time while minimizing unwanted side effects in the body. This involves the synthesis of various novel biocompatible polymers with well-defined nanometers to a few micro-meters structures using several modern techniques such as microelectromechanical systems [71] microfluidic systems [72-76], electrodropping system [77], microneedle based system [78-81], advanced high pressure homogenization, interfacial emulsion polymerization and combined systems. Figure 2 described the few modern techniques for polymeric nanoparticles preparation with various concepts. The physiochemical characters of polymeric nanoparticles have to be optimized based on the specific application. Various methods can be used to produce various nano-particulate systems with various polymers. The multifunctional polymeric nanoparticles developments such as environment-responsive micelles, colloids, nano hydrogel, core-shell nanoparticles, nano-spheres and core-shell nano-spheres with layer-by-layer assembly for single/dual or multi drug release have been achieved so far. In order to get the desired properties, the mechanism of formulation method plays a vital role. Thus, it is extremely beneficial to have synthesis mechanism at hand to approach multi-functional polymeric nanoparticles with exact physiochemical properties for a specific application.

The smart delivery systems of target bio-molecules have been concentrated of recent researches for various interventions. Particularly, various proteins, peptide, growth factors and cytokine therapy for various diseases play a vital role in regulating cellular responses, and thus the design of multi-functional polymeric particles delivery vehicles are closely associated with the regulation of multiple cellular events, likewise a wide variety of target bio-molecules have been investigated in numerous literature reports [82, 83]. Also numerous of delivery vehicles have been studied and reported recently, this chapter will cover same viewpoint and their brief discussion will be covered without attempting to all the work that has been done in this field. Various concepts are utilized in the design of delivery vehicles that are capable of ferrying multiple active ingredients in a self-controlled manner, with different release profile kinetics. The distinctive self-assembly of multifaceted nanostructures from an easy colloidal system has been of interest to design a material with distinctive characters for the use of drug delivery vehicles. The inter-and intra-molecular linkage via van der Waals interaction leads to dense-packed self-assembly periodic nanostructures. These structures could be colloidal particle or clusters, based on the assembly [84, 85]. The natural or semi-synthetic polymer-based self-assembled nanostructures have inherent capacity of the nano-carrier for delivering many kinds of active ingredients, because of good biocompatibility and degradation/resorption properties [86]. In the sonication methods (Figure 2a), the self-assembled nanoparticle was achieved by probe sonication, the process has been done by cavitation, nucleation and reversible locking concept, the formed nanostructure have more flexibility in the nature [87]. In this self-assembled and core-shell particulate delivery systems, including water-soluble polymeric drug compounds conjugates [88], block polymeric micelles [89-93], long-circulating polymeric micelles [94, 95], nano encapsulations [96, 97], and core-shell nano-spheres [98, 99] have been synthesized by in situ two-step semi-batch emulsion polymerization technique (Figure 2b), as vehicle to target suitable dose of drugs in an accurate and controlled manner. Also the core-shell nano-spheres have been achieved for pH-responsive controlled release, and delivery of hydrophobic anticancer agents for acidic tumor tissues [100]. Recently Choi DH, et al have optimized electrodropping system to produce a homogeneous biocompatible core shell capsules for angiogenesis in dual delivery system [77], and they particularly focused on regenerative medicine. This electro-dropping system can overcome from the particle aggregation and drug encapsulation efficiency (Figure 2d). Coming to the micro-fluidics, the recent science and advanced technology of manipulating micro/nano-scale volumes in micro-fluidic channels have significant impact on the various applications. Advances and inventions in micro-fluidics are awaited to enhance the preparation of polymer nanoparticles and shifting to clinical evaluation [101] most of the micro-fluidic systems for synthesis, polymer nanoparticles are still under development and they have the widest possible to develop because they are highly reproducible, easily modifiable and can be incorporated with other techniques [102]. Recently, various micro-fluidic systems provide rapid mixing without any stimulator, such as stirring or electric force; have been originated [103]. Among these various systems the flow-focusing [104], droplet mixers [105] are widely utilized and it enables micro-mixing within the micro channel [106]. The flow focusing squeezes the solvent stream between two anti-solvent streams, resulting in a rapid solvent exchange via diffusion take place (Figure 2c). The effectuation of these rapid mixing methods for the development of nanoparticles in continuous flow; the micro-fluidic system has been achieved the continuous flow, narrow sized, mono dispersed with high drug entrapment and better batch-to-batch uniformity in compared with conventional methods [107].

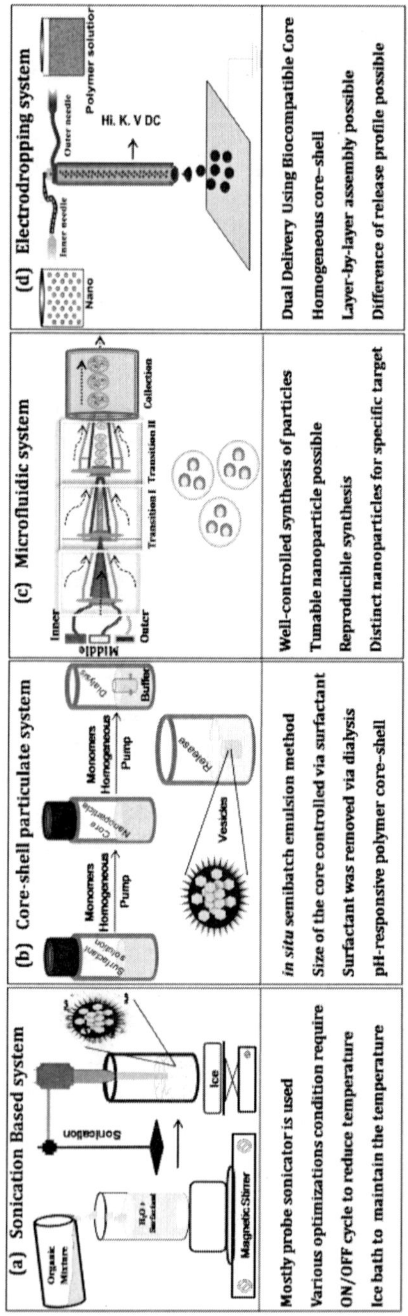

Figure 2. Schematic diagrams represent the advanced techniques of preparation of polymeric nanoparticles

4. CONTROLLED DRUG DELIVERY SYSTEMS

4.1. Rate-Programmed Drug Delivery Systems

The recent advances in smart drug delivery systems with rate-programmed drug delivery systems have been achieved by functionalization of rate-controlling surface. The transdermal drug delivery have been achieved a new rate pre-programmed drug delivery system, transdermal patch which delivers a particular concentration of drugs to the blood circulation via the skin, it provides the therapeutic advantage to clinical levels. The rate-programmed drug delivery systems, the release of drug molecules from the rate controlling membrane system has been pre-programmed at particular rate kinetics. The rate controlling membranes made from natural and semi-synthetic polymeric material and proves their ability to use as a rate controlling membranes in any dosage form even nano to micro-scale level particle embedded matrixes or implantable or transdermal patches. It must be simple, cost-effective, and flexible enough not to split or crack on bending or stretching. Recently, some of novel rate-controlling composite membranes have been developed as rate controlling barriers for transdermal application, with flexible and smooth surface nanoparticles embedded scaffold which could reduce the risk of wounding or being rubbed off during dressing, and thereby improves upon traditional dressings and its can provides better patient compliance [108, 109]. This is achieved by optimized system design, which determines the diffusivity of active agents across the membrane. This rate-programmed drug delivery system can be categorized by various controlling dependencies, such as (1): membrane permeation-controlled, (2): diffusion-controlled, (3): membrane/matrix hybrid-type and (4): reservoir partition-controlled systems. The recent advance in the smart rate-programmed drug delivery systems the polymer and their scaffolds play vital roles, such as greater drug-loaded nano/micro-particle encapsulation ability, overcome pre-systemic metabolism, enhanced bioavailability and environmental responsive properties for various applications. For selecting the polymers, need to consider some important key factors for pharmaceutical application such as reduced tensile strength [110], water vapor permeability rate, biocompatibility, non-toxic [111], anti-infective, controlled release [112, 113], flexibility, emollient, adhesion, spreadability and retention properties of the drug-loaded nano/micro-particle encapsulation scaffold or film preparation [114-116]. So it can prevent the immunogenesis, secondary damage to cells, disease recurrence and finally enhance patient compliance [117]. In this type of rate-programmed controlled drug delivery systems, a drug-loaded nano formulation or rate-controlled nano formulation can be either totally or partially loaded in the reservoir space whose surface is covered by the rate pre-programmed polymeric membrane. The pre-programmed polymeric membrane can be optimized and achieved by multi-functionalization with block copolymers. The scaffold or membrane can be produced by the homogeneous or heterogeneous non-porous polymeric compounds or a micro/nano-porous or semi-permeable material. The drug release profile should be at a constant pre-fixed rate. The release profile is controlled by a pre-programmed rate-controlling membrane; it's based on the molecules, diffusivity, partition co-efficient, and dimension of the outer membrane. Also the rate of release is determined by the cross-linking ratio of the polymer network. The rate controlled release profile exists in many kind therapeutic formulations such as intrauterine devices [118], ocular insert [119, 120], some transdermal therapeutic system [109], polymer matrix, sub-dermal [121] and subcutaneous implantation [122-125].

4.2. Activation-Modulated Drug Delivery

4.2.1. Environmental Activation/Stimuli Responsive Smart Delivery System

The smart drug delivery with activation-modulated system has been achieved by external or environmental stimuli, these environmental responsive smart delivery systems achieved a lot more with double and multiple-responsive delivery system. The various activation/stimuli responsive drug delivery vehicles have been synthesized and tested, in various particle sizes, ranges from nanometers to a few micro-meters sized carriers for different routes of administration. The transdermal electro-activated or electro-modulated drug delivery has been established as an efficient model. In this group of activation-modulated controlled drug delivery system, the release of active agents from the systems is activated by some physical, chemical, electrical, environmental condition or biochemical processes and/or facilitated by an energy supplied externally. The release profile has been controlled by the input energy. Based on the activation/stimulation process applied or energy type used, this activation-modulated controlled drug delivery system can be categorized into the various classes which are given in the Table 1. These stimuli-responsive materials show changes in the physicochemical character during the environmental condition changes. These changing properties can be fully utilized in smart delivery system, which certainly similar to the biological response behavior. Different types of body organs, different tissues and various types of cellular compartments might have great differences in every stimulus with great response. So that all the important cases considered in this chapter, deal with various environmental responsive smart delivery systems. Any specific behavioral changes in the system lead to a phase transition, these transitions will be key factors for the stimuli-responsive drug delivery system and some selected examples of applications are described in the Figure 4. The preclinical and clinical studies have demonstrated that drug-loaded polymeric nanoparticles has been well tolerated, extended systemic circulation, higher accumulation in the tumor sites through enhanced permeability and retention effect, minimized side effects and adverse effect, and/or higher bioavailability [153-155]. And most of the drug delivery systems are based on biodegradable polymer [156, 157]. Most of the environment-sensitive polymeric nano-particulate systems are leading to degradation and or disintegration by the internal or external local environmental stimulus such as pH, glucose, low oxygen content, ions, redox potential, and lysosomal enzymes; and then temperature, magnetic field, electric, ultrasound, and light respectively (Table 1).

These activations grew to achieve smart, targeted drug release in a particular time (spatial and temporal control release) [158-160]. At this place we describe a few examples. Particularly, the acidic pH levels in the body vary according to the different body environments (site and the organ) such as tumor cells and tissues (pH 6.5-7.2), endosomes (pH 5.0-6.5), lysosomes (pH 4.5-5.0) and entire GI tract with different pH value as comparatively varied with normal physiological (pH of 7.4) conditions in blood and tissues. So, the pH-responsive nano system have been considered and formulated to release the active agents in pH sensitive targets such as cancer site or endo/lysosomal regions [161,162]. The cytosol and cell nuclei have surrounded with elevated redox potential (in reducing glutathione) it higher than normal body fluids and it have been developed for intracellular release of various active bio-molecules [163-165]. Additionally, the cancerous tissues are extremely low in oxygen content (hypoxia) with higher glutathione levels compared to normal tissues [166]. This has been targeted with hypoxia-responsive polymeric nanoparticles. These internal stimuli-responsive nanoparticles have their own benefit of self-regulated drug delivery and effective target in clinical therapeutics. Also the external activated nanoparticles provide their own advantages such as high reproducible nature, also remote controlled delivery possible, then the release profile can be pulsatile delivered (means that switched on and off) possible [167]. On the other hand, the various light-responsive polymeric nanoparticles system has been developed for activating antitumor drug release [168]. Also numerous of temperature-sensitive multi-functionalized polymeric and copolymers nanoparticles have been formulated based on thermally-responsive release [169, 170]. Magnetically guided nano-carriers have been developed for the remote controlled cancer therapy and diagnosis [171, 172]; also the core-shell nanoparticles have demonstrated for improved tumor accumulation and antitumor therapeutic efficacy in various models.

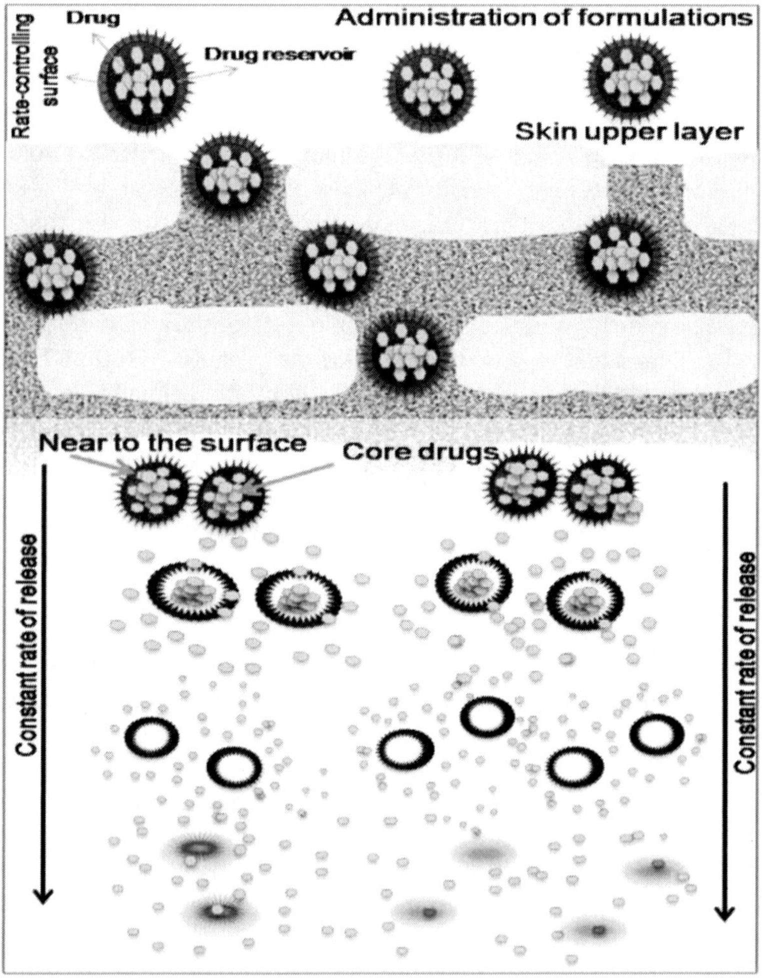

Figure 3. Schematic diagrams represent the rate controlled drug delivery systems of topical applications

Table 1. Overview of various stimuli responsive nano-carriers for smart drug delivery systems with mode of drug release applications

Based on	Stimulus	Mode	Ref.
	Hypoxia	Hydrophobically modified imidazole derivative was conjugated to the carboxymethyl dextran, it can release the hydrophobic agents under hypoxic conditions	[148]
Dual-stimuli	Two different responses	Based on the polymer architecture	[149]
		Micelles are reported pH and thermo-responsive	[150]
Multi-stimuli	More than two responses	Functionalization of pyrene-quaternized segments form a light-responsive shell and the unquaternized segments form a temperature/pH-responsive core	[151] [152]

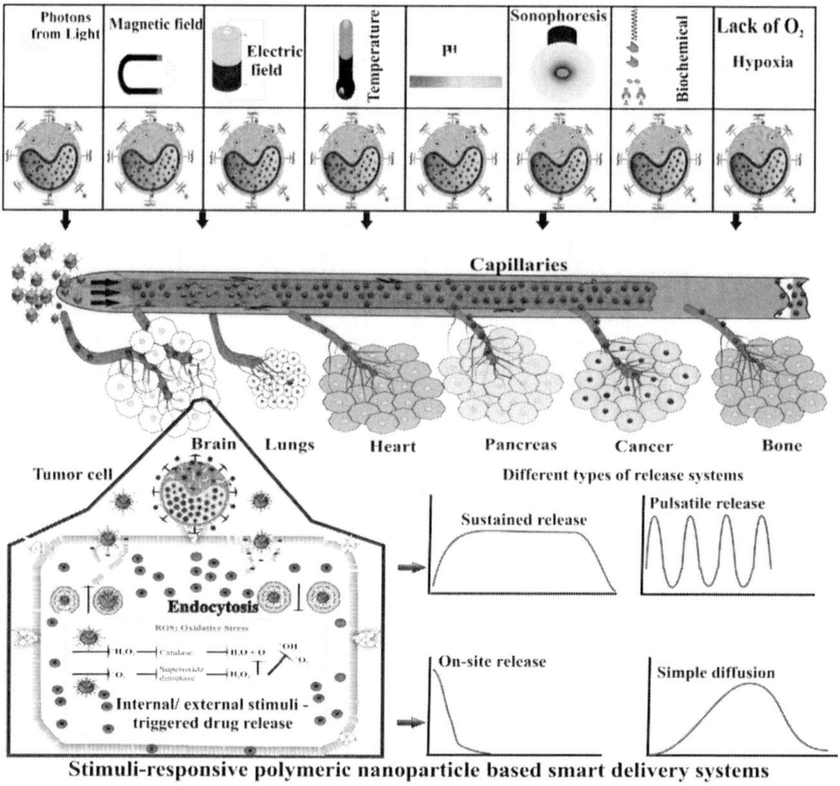

Figure 4. Schematic diagrams represent the activation-modulated drug delivery systems, which the polymeric nanoparticle activated by various stimuli such as physical, chemical, biochemical, environment, and/or a combination of two or more.

4.2.2. Dual and Multi-Stimuli Responsive Smart Delivery System

In this chapter, provide the recent proposes and formulations of dual and multiple-stimuli responsive multi-functionalized polymeric nanoparticles and their promising targets in smart drug delivery in specific to the cancer therapy. With the booster development of the smart drug release and increase therapeutic efficiency of intelligent drug loaded nano-particulate system, polymeric nanoparticles that respond to dual and multi-stimuli, which have been aggressively reported. The double-response and multiple-responsive nano-particulate systems were described in the Table 2. It must be mentioned that the stimuli and responses happened at the same time at the same site or different mode. These dual and multi-stimuli responsive polymeric nanoparticles can provide control over the drug release profile, which leads to greater anti-tumor efficiency in vitro and in vivo models, and on the other side the nanoparticle formulation and drug loading under moderate conditions. In this section we describe a few examples. Especially, redox-responsive drug release multi-functionalized nano-particulate system have been formulated based on temperature and reduction, dual responsive tri-block copolymers functionalized by increasing temperature above the lower critical solution temperature after that cross-linking [173, 174]. These multi-functionalized nano-particulate systems were targeted to cancer cells and triggered by reduction oxidation mechanism, which leads to dissociate to release the active agents by de-crosslinking followed by disruption and degradation of nano-particulate system. pH/redox dual-stimuli multi-functionalized disulfide cross-linked micelles have been developed for increased drug release and accumulation in the cancer target, due to endo/lysosomal pH and intracellular redox environment the drug release was taken place [175].

4.2.3. Considerations For Stimuli Responsive Targeted Molecular Systems

These multi-functional polymeric nanoparticles are capable to face the current problems of nanoparticle drug formulations including formulation and drug encapsulation, prolong stability, cellular internalization, site-targetability, enhanced cellular uptake, and inside cell target and drug release. These dual and multiple-activation responsive characteristics have provided novel and enthusiastic power over drug release kinetics and greater efficiency. All the described studies in dual and multiple-stimuli responsive drug delivery systems are mostly trial and error models, because most them non-biodegradable carriers, low encapsulation, and nonviable to clinical therapeutics. To overcome all the unfavorable conditions, immediate efforts could be focussed to improvement of dual and multiple-stimuli responsive biocompatible, biodegradable, non-toxic, and non-immunogenic smart polymeric nanoparticles that could effectively entrap and sustain the drug release in the systemic circulation, enhanced accumulation in the cancer target, and efficient release kinetics in response to more efficient external or internal stimuli. Moreover the smart polymeric nanoparticle system does not produce any secondary damage and any harmful to the healthy cells. In the case of clinical studies on dual and multiple stimuli responsive system shall be performed to obtain a real mechanism of action in anti-cancer target. In addition, the multi-functionalized smart polymeric nanoparticles system construct with targeting ligands and shall be incorporated into dual/multiple stimuli responsive nanoparticles to be achieved multidrug resistant cancers by site targeting, site-specific, and rapid/sustained release, and we sure that dual and multiple stimuli responsive smart nano-particulate system going to be a good future in cancer therapy.

Table 2. Overview of dual and multi-stimuli responsive materials for nano-carriers of various smart drug delivery systems

Responses	Stimulus	Nanoparticles	Ref.
Dual-stimuli	pH & Thermo	P(NIPAAm-co-DMAAm-co-UA) nanoparticles	[176]
		P(NIPAAm-co-AA)-b-PCL nanoparticles	[177]
		PLA-g-P(NIPAAm-co-MAA) nanoparticles	[178]
		P(NIPAAm-co-DMAAm)-b-PCL/PLA micelles	[179]
		PNIPAAm and PAA hollow nanogels	[180]
	pH & redox	PEG-SS-PDEA polymersomes	[181]
		DS-g-PEG/cRGD nanoparticles	[182]
		Poly(b-amino ester)s-PEG micelles	[183]
		PMAA-based nanogels	[184]
		mPEG-PAsp(MEA)-PAsp(DIP) micelles	[185]
	pH & magnetic	Fe_3O_4 nanocarrier with peptide mimic polymers	[186]
		DOX-tethered Fe3O4 conjugates nanoparticles	[187]
		mPEG-b-PMAA-b-PGMA-Fe_3O_4 nanoparticles	[188]
		Fe_3O_4-capped MSNs	[189]
		MCM-TAA-Fe_3O_4-capped MSNs	
	T & redox	EO-PAA-PNIPAAm polymersomes	[190] [191]
	Double pH	PPC-Hyd-DOX-DA nanoparticles	[192]
		Poly-b-amino ester ketal nanoparticles	[193]
	pH & diols	PEG-b-dendritic cholic acid telodendrimers nano-carriers containing boronic acid	[194]
	T & magnetic	Pluronic with Fe_3O_4 nanoparticles	[195]
	T & enzyme	DNA-capped MSNs	[196]
Multi- stimuli	T/pH/redox	PNIPAAm-SS-P(THP-protected HEMA) micelles	[197]
	T/pH/magnetic	P(NIPAAm-co-MAA) coated magnetic MSNs	[198]
	pH/redox/ magnetic	Fe(II) loaded PMAA crosslinked by N,N-methylene-bisacrylamide and N,N-bis(acryloyl)- cystamine	[199]
	T/redox/guest molecule	Vesicles based on hostequest complex formation between C4AS and MVC12	[200]
	T/pH/guest molecule	Cucurbit(8)uril micelles, methylviologene-functionalize PNIPAAm and naphthalene-terminated PDMAEMA	[201]
	Light/pH/T	Pyrene-functionalized poly (dimethylaminoethyl methacrylate)	

4.3. Feedback-Regulated Drug Delivery

The recent advances in smart delivery systems with feedback-regulation of drug release. This self-regulated or feedback-controlled drug delivery comes under closed-loop systems. The self-regulated system drug release rate is controlled by feedback information, without any external stimulation, and utilized several approaches to control the release rate [202-205]. The feedback-regulated drug delivery concepts were schematically depicted in Figure 5. The feedback-regulated drug delivery concept has been applied to the development of various controlled delivery systems such as bio-erosion regulated, bio-responsive regulated and self-regulating drug delivery systems. Among this one of the concepts has been involved in the smart controlled delivery systems. For that various research efforts are also in progress to develop such nanoparticles that contain drugs capable of a feedback-modulated drug release. The drug release is activated by a triggering agent, such as a biochemical substance, in the body via some feedback mechanisms. The release rate has been determined by triggering agent concentration. When the triggering agent is above a certain level, the release is activated. This can induce and stop the drug release. It would be a high potential benefits if they were delivered by a system that recognized the particular warning signal caused by disease affected part, then they estimated the magnitude ratio of the signal, and then acted to release the exact quantity of active drugs in response. This kind of drug delivery system required to fulfil the physiological need by means of some feedback mechanism. The self-regulated drug delivery systems utilize several approaches for the rate-control release: pH-responsive polymers, temperature-responsive polymers, enzyme-substrate reactions, antibody interactions, enzyme-mediated, pH-dependent drug solubility nature, competitive binding mechanism and metal concentration-dependent hydrolysis. A hydrogel can swell in aqueous medium and retain their structure. The multi-functionalized polymer nanoparticle can be incorporated into hydrogel, such hydrogel used for the feedback-regulated drug delivery system. This hydrogels can protect the drug from dangerous environments such as enzymes and low pH in the stomach. This can control drug release through changing the network structure in response to particular stimuli, which can enable the sensor leads to reversible volume phase transitions upon small changes in the environment condition. For example, the polymers characterized by lower critical solution temperatures generally shrink, as the temperature is increased via lower critical solution temperature. Decreasing the temperature below lower critical solution temperature, the polymer can swell. Biomolecules can be encapsulated on or within the heat responsive polymers.

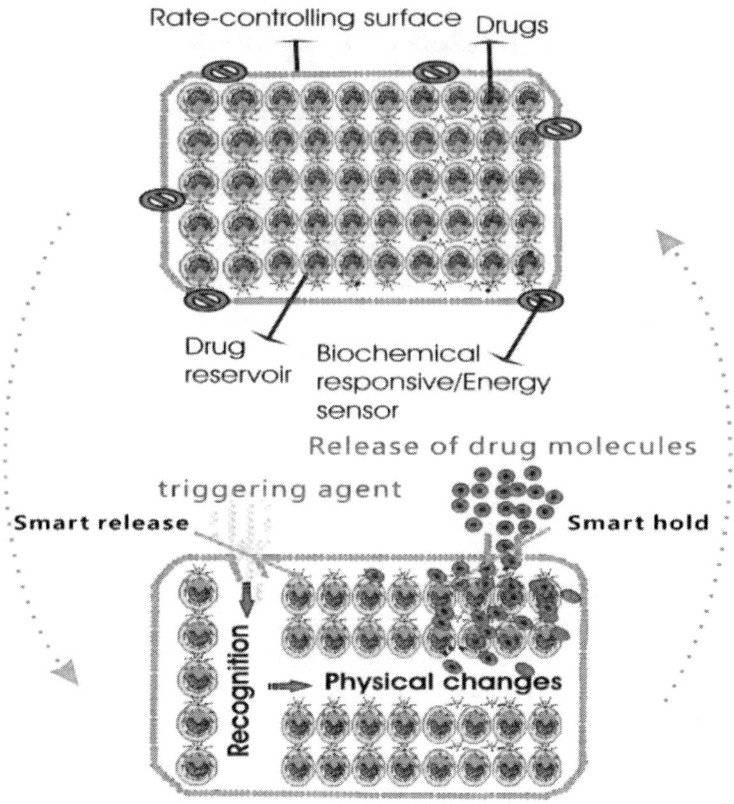

Figure 5. Schematic diagrams represent the feedback-regulated drug delivery systems

The sensor grafted in the delivery system can enable to mimic the recognition function of various bio-chemicals such as enzymes, cell mediated receptors and various proteins in human beings for maintaining the regulation and equilibrium. This approach is utilized for drug incorporated polymeric feedback controlled delivery systems, and this system approach is based on the observation that changes in control mechanisms, e.g.: pH or ionic strength or temperatures can affect large changes in drug solubility; this can be the main factor for control release rate. The external trigger molecule and polymer-bound enzyme can alter the pH inside the polymeric system. If the pH alteration happened inside the polymer system that can lead to changes in drug solubility, which is induces the diffusion or dissolution or disintegration, and rate of release has been changed accordingly. Many researchers have been developed a membrane to bypass the rumen but it allows the polymeric system to release the drug in the stomach via gastric retention mechanism [206]. Because of the polymer membrane it is impermeable to the rumen pH 7, but the swells and release at pH 4, which is the fourth stomach. Several studies have been performed on various polymers holding weakly acidic or basic functional groups in the polymeric backbone [207-112]. This polymeric system can swell or de-swell by changing the pH of the environments. By this way the drug will release from a matrix or device, which is developed by pH dependent polymers and this system can provides controlled release rates.

The bio-erosion controlled drug delivery system comprises of a drug-encapsulated bio-erodible scaffolds developed from biocompatible polymers (poly (vinyl methyl ether)), and were layered using immobilized urease. In a neutral pH the polymer erodes gradually, but in existing with urea, urea is metabolized by the system containing urea to form ammonia, it leads to increase the pH in the surrounding area, this increased pH degrade the polymer scaffolds then the drugs has been released [213], and some polymers require high pH to degrade.

The bio-responsive controlled drug delivery system, glucose-triggered insulin delivery has developed [214], the insulin is encapsulated within biocompatible polymer hydrogel scaffold comprising abundant NR2 functional groups present in the normal state. So in this state scaffolds are un-swollen and thus impermeable to insulin molecules. Enzymatically oxidized glucose is to form gluconic acid, this triggers the NR2 groups to form NR2 H+, it leads to swollen and insulin molecules deliver through the polymer membrane, and the amount of delivery has been controlled by glucose penetrating concentration.

The reversible and competitive binding mechanism also has been reported to insulin delivery. This mechanism role is to activate and to regulate the release of drug in the target; also it depends upon the glucose level present in the systemic circulation. Insulin-sugar-lectin complex has been prepared and entrapped into the semi-permeable polymeric membrane to achieve controlled release. The diffused blood glucose has competitively bound to particular binding sites, then activates the complex to release insulin derivatives, and the release acted based on the concentration of glucose presented in the systemic circulation. By this way the self-controlled drug delivery has been achieved. A further improvement on insulin delivery, they used glycosylated insulin-concanavalin A complex and entrapped inside polymeric membrane and the release has been achieved by self-regulated mechanism, depends on the glucose concentration permeate into the system [215]. Again in the development of self-regulating insulin delivery has achieved by enzymatically controlled implantable glucose-dependent insulin delivery systems [216]. Followed by various researches developed the different kinds of glucose-responsive insulin delivery [217-223]. Also the molecular imprinting technology developed system able to identify the specific compounds on the cell surface, and this can be appropriate for further developing and targeting the delivery system to specific tissues or cells. Recently, the pH-Sensitive polymer multi-functionalized with block co-polymeric nanoparticles have been developed for the triggered release of paclitaxel within a tumor microenvironment which the polymer acted as a feedback-regulated drug delivery carrier [224], and this carrier have a reversed swelling behavior. Most recently, the feedback controlled drug delivery system has been developed for cerebral cortical disorders with a feedback controlled mechanism. Drugs have been delivered via subdural/subarachnoid space, then diffuse into neocortical tissue and this diffusion can be controlled by electrophysiological feedback, the cerebral cortical area is exposed to the drug, and they were optimized for the drug concentration, delivery, frequency of delivery [225]. Moreover, the molecular imprinting technology has a huge possibility for producing acceptable dosage forms in the feedback-regulated drug delivery systems. The application of molecular imprinting enables the design of new systems and also in polymer based device fabrications. The advances in the preparation of molecular imprinting as spherical uniform particles [226] and scaffolds [227] can increase the field application potentiality of several polymers in drug delivery system. Moreover, these imprinted delivery systems have not yet touched in clinical therapeutics.

4.4. Site-Targeting Drug Delivery Systems

The recent advances in the smart delivery systems with site-targeting drug release. A site targeted drug delivery systems are complex of multiple steps of diffusion and partitioning. Nowadays the site targeted drug delivery systems involve deep investigation as they are very eager to overcome the modern medical application [228]. A well-designed multi-functionalized polymeric carrier for site-targeted drug delivery in the interventions of various diseases such as colon disease, kidney/renal disease, nasal disease and genitourinary disease has been reported recently [229-234]. A variety of both natural and synthetic water-soluble polymers have been used for biomedical applications. These polymers have been used routinely in biopharmaceutics because of the effectiveness in controlled drug release. The traditional formulations are not significantly efficient at targeting molecules, thus the new and smart drug delivery systems are being studied to overcome the problem. The goal of the smart drug delivery systems is to allow a localized drug delivery, at the same time; it does not affect the healthy tissues and no unwanted effects. The drugs composed of micro-or nano-sized particulate system, which is able to spread through the systemic circulation, and transport through various body organs and body areas such as arteries, veins, and capillaries and even cross membrane barriers. The nanoparticle transport and targeting tissue are the complex process, so the transportation and communication have been viewed by the molecular communication paradigm. This transport of drug-loaded particles in the human body has been viewed, where the nanoparticle has transported this information is conveyed by signaling molecule. This communication system provides a clear reading of particle diffusion, distribution, disintegration over time throughout the biological system, which provides the importance to the invention of a smart particulate delivery system. Initially, the kinetic Monte Carlo method [235, 236], have been used computer simulation to solve the communication system. Lately, researchers developed an analytical approach based on the abstraction of targeted particulate delivery systems as a communication mechanism. This information is passed between sender and receiver by intracellular and intercellular signalling [237]. Different kinds of molecular communication have been analyzed so far, which involve passive or active transport of molecules [238,239]). The smart site targeted delivery system takes an advantage of the systemic circulation for the distribution of active drug particle from where it's ingested to the systemic circulation to a targeted site. Basically, the delivery systems have been made with purpose and intention to control the rate of release from the systems, but the transport of nanoparticle to the target site still needs more control. Preferably, the route of administration and nanoparticle transport should also be strong enough controlled.

In this section also provides a few examples of site-targeted drug delivery systems, The ideal example is that the kidney site-targeted drug delivery systems, it acted as a smart delivery to enhance drug efficacy and safety in the therapeutics of kidney diseases. By this smart drug delivery treatment provides that reduces inflammation and reduce the formation of excess fibrous to proximal tubular cells, it can protect systemic infection and renal tubular inflammations. So targeting the renal proximal tubular cells is the novel and efficient routes to cure kidney disease [240-244]. Kidney-targeted drug delivery system can overcome from the various obstacles such as kidney transplantation, ureteral obstruction, diabetes, and other some important kidney disease. Figure 6 shows the kidney drug delivery of nano-particulate systems. Among all drug carriers the macromolecular carriers are extremely powerful targeting the kidney, because of the selective accumulation in the kidneys. Macromolecular carriers with prodrugs play crucial roles in targeting drugs to particular target cells in the kidney. The molecular weight and electric charge of polymers is one of the crucial role for effective renal clearance [245, 246], thus the active polymeric system can uptake and exists in the renal cells [247]. Especially the multi-functionalized polymeric nanoparticles showed higher uptake in glomerular mesangial cells [248, 249]. For nasal site-targeting specificity, the multi-functional particulate system design is the main role for site-targeting. So, design and preparation method has to be controlled according to the needs, the materials should be with quality of properties such as biocompatible, biodegradable, modifiable, mucoadhesive, antimicrobial, tumor or particular cell recognition, and maintain the drug release. In the example, N,N,N-Trimethyl chitosan nanoparticles achieved controlled intra nasal delivery to treat various diseases including hepatitis B and allergic rhinitis [250]. Also the amine functionalized chitosan has been shown their eminent characters such as biocompatible, enhanced solubility, strength, porosity, absorption efficacy, chemical tolerance, non-immunogenic and non-antigenic properties, and it has been used for various nasal delivery.

4.5. Targeting Strategies For Kidney Diseases

Macromolecule is a very large molecule, which can accumulate in the kidneys. Generally, the molecular weight of the macromolecular vehicle is bigger than that of the prodrugs, so this kind of system can achieve the goal. Pro-drugs have the ability to select the target in the kidney because it can release the active drug by the action of renal enzymes. The various strategies of kidney-targeted drug delivery systems has to be considered such as biodynamical strategy of renal artery perfusion, macromolecular carriers which includes enzymes, immune proteins and peptide hormones, pro-drugs which includes folate, sugars, and amino acids, and other strategies including various nano-particulate systems. The molecular weight and charge [245, 246] of polymers is the main factor, it can influence their distribution in various organs including kidney. In general, increasing the molecular weight of polymers leads to decreases urinary clearance. Some of the polymers have been eliminated rapidly from the systemic circulation but it does not excrete from the kidney, and its accumulated in the renal systems. So it clearly proposed that the selection of effective and active multi-functionalized polymeric nanoparticles can uptake by the particular kidney cell types. So the selection of polymers is one of the prime strategies for consideration to achieve the efficient kidney targeting. These new possibilities to develop kidney targeting conjugates and other nano-particulate drug delivery systems. Including various polymers based nanoparticles give excellence strategies to achieve the goal of targeting drugs to the various renal diseases.

4.6. Common Strategies For Smart Polymeric Particulate Targeted Delivery

The ideal proposed model for site-targeting delivery is fabricated from a biocompatible, non-immunogenic and biodegradable polymer and acts as the central of support to three main characteristics of attachments such as site-specific targeting moiety, solubilizer and drug moiety, which should have drug delivery capacity, capable of transport and active molecule should bonded to the polymer via spacer, and the linkage is cleaved by particular enzyme(s) at the final targeted site respectively. In order to develop a new polymeric vehicle for a particular drug, the polymer distribution in the systemic circulation has to be analyzed since it's right away affects on activity of drugs. For controlling the systemic distribution of drugs, we need to consider minimum two strategies which are active or passive targeting. Previously, the drug is delivered to target site using some specific antibodies, which are specific to target cell-surface [251-254]. This method gives efficient targeting to tumor site; however, the antibodies can produce immunogenic activity. But, the passive targeting with bio-polymers vehicles cannot produce immunogenicity or toxicity, this might enhance the active molecule efficacy, such as increased half-life by increased size of the nano-particulate complex, increased permeability at the targeted area and polymer vehicle interacts to the body organs. Those elements must be increase the absorption of the drug molecule; which minimize the dosage and low unwanted effects [255, 256]. Moreover, in the advanced fabrication of molecular imprinting technology can provide efficient smart polymeric systems with the ability to recognize specific bio active molecules. This advanced fabrication technology has tremendous possibility to meet the requirements for satisfactory dosage forms developments. Depends upon the particular application the fabricated systems can decide the delivery, efficiency, safety of the drugs, and when it should be reached.

Described all above application strategies have a significant interest in targeting drugs into specific regions [257, 258].

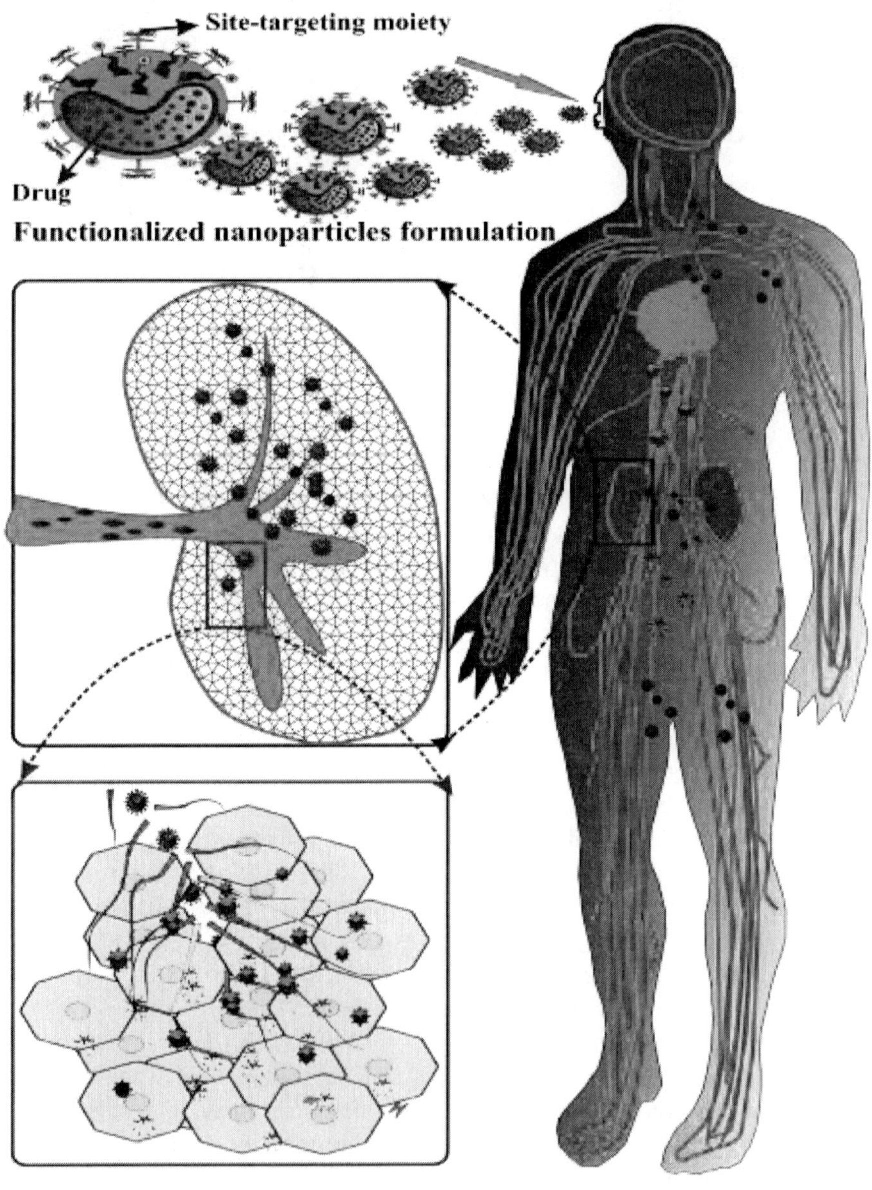

Figure 6. Schematic diagrams represent the site-targeting specificity particulate drug delivery systems

5. BIOENGINEERED MATERIALS: IDEAL AND RECENT ADVANCES FOR DRUG DELIVERY SYSTEMS

5.1. Nano-Engines Of Drug Delivery Systems

Engineered materials have been utilized for developing smart drug delivery systems. Design and multi-functionalities fabricate of efficient smart drug delivery systems are vitally necessary for medicine and healthcare development. In the material science field provides biodegradable, biocompatible, environment-responsive, and highly effective novel polymeric system for targeted delivery. Nanotechnology provides bottom-up and top-down nanofabrication with size controlled and multi-functionality of particulate for targeted delivery. New materials invention and advanced technology have been synergistically achieved in drug delivery so far. The essential goals of medical pharmacology provide the right medicine, right dosage, and right route at the right time to the right patient, so more research need to optimize the therapeutic efficacy of the drug. This is the essential principles is behind the smart drug delivery. A smart, controlled delivery system needs synergistic consideration of several factors; these have been summarized in Figure. 7. It is difficult to get all consideration factors in a smart controlled delivery system due to other influencing factors. Also high quality, reliability, efficiency and reproducibility are the most significant issue while designing such a smart system. Also the smart systems have to induce the drug release and stop the release by their own manner. It would be highly benefited, if the system recognizes the disease affected part, estimated the disease affected ratio, and then acted to release the exact quantity of active drugs. This kind of drug delivery system can fulfil the medicine and healthcare requirements.

6. POLYMERIC NANOPARTICLES FUNCTIONALIZATION AND CONSIDERATIONS FOR SMART APPLICATION

In this section provides the recent research on the preparation and functionalization of various polymeric hybrid nano-materials including nanoparticles and microparticles by various techniques. Several techniques have been developed for the functionalization of polymeric nanoparticles with different therapeutic applications. The polymeric nanoparticles have been studied for their enriched properties in biological systems, with the nature of the materials and whether it has the specific properties for chemical modification and functionalization of the nanoparticle developed from various materials including bio-macromolecules. There are several researchers have been studied for functionalization and surface modification of nanoparticles and it would not cover all this in this section; so, this section covers some examples of nanoparticles functionalization and some important criteria to consider the fabrication process. In addition, the richness of surface chemistry and potential biomedical applications are described. The polymeric nanoparticles surface functionalization are mainly two types, one is functionalization with biological (macro)molecules such as peptides, carbohydrates, lipids, fatty acids, proteins, and nucleic acids (genes, oligomers, aptamers, and ribozymes/DNAzymes); another one is functionalization with specific ligands such as mono-or oligosaccharides (carbohydrates), folate receptor, antibodies and biotin are commonly used. This surface functionalization have been made by various modifications on preformed nanoparticles through adsorption, functional surfactants, emulsification, polymerization, covalently bounded functional molecules and various forms of bio-conjugation. There are few considerations for functionalization of polymeric nanoparticles properties such as: 1) the bio-molecule ratio should controlled by calculating the number of conjugate sites presents in the nanoparticles with different applications, 2) due to the environment and electrostatic interactions the alignment of functionalization has been varying, so the non specific attachment should be avoided in the performed nanoparticles, 3) depends on the applications requirements the nanoparticles bio-molecule distance should be maintained, 4) control the conjugation moiety attachment/linking affinity to the performed nanoparticles, 5) should maintain the optimal efficiency of physiochemical characters and 6) it should be high reproducible for all batches. The above all criteria can fulfil the requirement of design and functionalization of nanoparticles for a controllable release profile that satisfies the desired application. And better protection against environmental factors and maximum optimal control is achieved if drug loading is carried out by encapsulation instead of adsorption on to the particle surface. With the combinations of these above criteria in the fabrication of nanoparticles are potential to increase the clinical therapeutics by reducing unwanted effects.

With the field of bio-nanotechnology, enormous new research on the synthesis of polymeric nanoparticle based top-down or bottom-up approaches have been recently developed. Recent developed polymeric systems engrafted nanoparticles provide the optimal characteristic of the functionalized nanoparticles for various therapeutic approaches in harsh environments such as in the acidic and alkali environment [259]. Also polymer nanoparticles are broadly used in several therapeutic applications, mostly cancer targeting and therapeutics. And we provide some examples of various nanoparticles with different functionalization and different therapeutic uses based on the target, shown in Table 3. Therefore, the multi-functionalized nanoparticle over comes from the drawbacks of conventional therapy. In the latest study provided that more than 26 nanoparticle based therapeutic system have been approved for clinical treatment and several nanoparticles are under consideration [281]. In order to achieve the efficient nano-particulate system based therapeutics the nanoparticle synthesis and functionalization methods have to consider very carefully. Although several surface modified methods for various bio-applications have been reported previously, in this section highlight particular examples where this type of functionalization has been used.

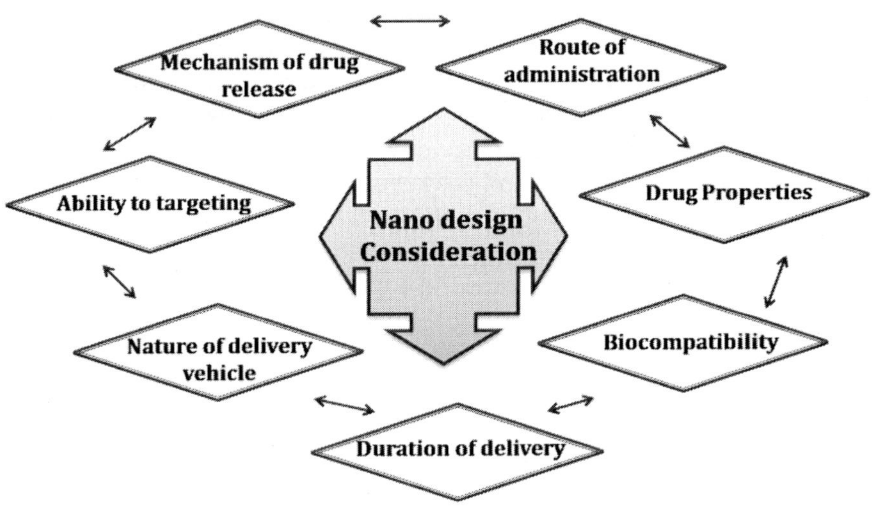

Figure 7. Requirements of several factors for simultaneous consideration to design a polymeric nanoparticle for the smart drug delivery system

Functionalization is defined as the improving performance of nanoparticle by a chemical functional group on their surface. Some basic components of functionalized nanoparticle are enabling to increasing the multifunctional applications in the field of biomedicine; the basic components are diagnostic agent, targeting ligand, spacer group, therapeutic agents, and polymer nano-carrier with proper functionalization. Here we introducing two strategies for surface functionalization, first one is direct functionalization, where the functional ligand is a bi-functional compound. In this method, one of the reactive groups is used to bind to the nanoparticle surface and the second group contains the required active functionality. Another one is post-functionalization, here the strategy is not changeable and the nature of the functionalizing group cannot be compatible with good control over the size and dispersion of the nanoparticles in the solvent used for the fabrication. Commonly, the nano-carriers have been functionalized with various chemical functional groups such as thiols, disulfides, amines, nitriles, carboxylic acids, phosphines and bio-macromolecules [282-287], based on their application. The functionalization of nanoparticle is to modify their outer surface with other specific chemical agents based on the desired application. After functionalization the particle physiochemical character has been changed. Also, it is a very important step for control because it can change their size and self-organization during the formation and should not promote aggregation. The prepared polymeric nanoparticles have emerged promising technology platform for recognizing the target with navigated controlled drug delivery system. Figure 8 shows the various functionalizations of the nano-engines for the development of smart drug delivery systems (Left side) and pre-regulated nanoparticle recognizes the tumor cells not the healthy (right side). This therapeutic drug concentration reaches the tumor site not in the normal cells or tissues. Polymer base smart drug delivery can overcome the patient complaints in healthcare.

In polymeric based nano-composites fabrication, the nanoparticles is used as backbone to enhance the physiochemical characters [288-290] such as flexibility, smoothness, enough strength and stiffness, which are much essential in the field of tissue engineering and bio-medical applications. The mechanical strength of polymer based nanocomposites is low due to the poor linkage between nanoparticles and the polymer, which leads to artificial defects in the composites [291-293]. It could be engineered with the appropriate interface to enhance the flexibility, smoothness, strength, stiffness and compatibility of the composite character [294]. The advanced functionalization of the nanocomposite have been prepared with suitable surface active agents, including anionic and non-ionic surfactants, it can lead to strong linkage between the nanoparticle and the polymer. The multi-functionalized nanocomposite enhances the physicochemical properties and no untoward effect on the biological system had been reported [295]. For the hydrophobic drug the phage display technique has been used for the functionalization [296], and the bioavailability have enhanced by post-polymerization. Additionally, the post-polymerization with copolymer produces efficient targeting in the extracellular compartment of the biological system [297-300]. With the nanoparticles the polymers like PEG establishes for prolonged systemic circulation [301, 302]. For the stimuli responsive targeted drug delivery has been achieved by the functionalization of suitable materials (light or magnetic or thermal or ionic responsive material). Particularly, the magnetic induction systems have been used with functionalized magnetic nanoparticles for cell or tissue specific targeted delivery. For targeting brain delivery system the nanoparticles has been functionalized for specific or nonspecific binding mechanisms [303]. The fabrication and functionalization science has merged with software oriented technology for the development of controlled and targeted nanoparticle loaded micro-device system [304]. The recent trends in novel polymer and block co-polymer synthesis methods like radical polymerization and click chemistry has been provide well-desired multi functionality polymeric structures [305-312]. This is the potential method to fabricate the desired molecular weight polymer with well-defined characteristic features. This unique method of polymer synthesis gives the successful nano formulation for potential bio-application. The functionalized nanoparticles have been synthesized with potential biochemical moieties. Then these multi-functionalized nanoparticles have been examined for desired physicochemical property and biocompatibility.

7. RECENT DEVELOPMENTS, SIGNIFICANT ROUTE OF ADMINISTRATION AND TARGETING STRATEGIES

The route of administration of therapeutics is crucially important to cure the disease. Despite the invention of potential therapeutic moieties, the inefficient drug targeting by pills or injection on the appropriate site of the body limits therapeutics values to a larger extend. There are multiple barriers involve in the anatomical and physiological system to lack the drug efficiency, including enzymatic degradation in the stomach, absorption across the intestinal epithelium, hepatic clearance, and accumulation in non-targeted tissues. These barriers also involve a range of complexities from the tissue to the organelle level along with the time that mismatch the drug potency in vivo. Collectively, these conditions challenge the active utilization of potent therapeutic molecules for disease treatment or prevention. Extensive research has been carried out in the field of drug delivery to overcome these challenges and thus to contribute a significant role in the overall drug-development process. After the evolution of nanotechnology and vast increment in knowledge about the human body, advances have been achieved in the drug delivery field as targeted delivery and sustained/controlled delivery system. By tuning the kinetic properties of therapeutics, the potentiality could be secured until it reaching the targeted organ and this factor is considered to be the most important in the field of pharmacology. Progresses in the nanomaterials development have been fruitful to fulfil the goals of drug delivery. Pharmacologically, the drug delivery is better explained based on the routes of drug administrations. Development of alternative drug delivery methods is crucially important to overcome the challenges experienced throughout the history of medicine. Scientists have been working on the creation of the smart drug delivery system and such approaches could provide an easy route of administration, ensuring patient compliance, decreasing toxicity, improving bioavailability and achieving precise therapeutic targeting. Creation of smart drug carrier as delivery systems and the discovery of new pharmacological compounds will potentially advance disease diagnosis and treatment beyond expectation. A variety of novel drug delivery systems have been developed using various nanomaterials during the last decade and several of them are already marketed. Nanotechnology manipulates the multiple properties including the size and other physical characteristics and thus achieves both controlled and targeted delivery of drugs. The bio-adaptability and multi-functional properties of smart delivery system minimize the undesirable properties of drugs in various routes of administration, including oral, rectal, nasal, ocular, topical route such as transdermal, and dermal, parenteral route such as intravenous/intravascular, intramuscular, subcutaneous, intradermal/intracutaneous, intraperitoneal and intrathecal. Figure 9 depicts the tremendous applications of new nanomaterials for the development of various routes of administration and targeting for therapeutics such as transdermal vaccine delivery, intranasal vaccine delivery and lung targeted delivery. Nasal mucosa offers numerous benefits as a target tissue for drug delivery, particularly for brain targeting because drug penetration through the BBB is favored by lipophilicity.

Table 3. Examples of various nanoparticles with different functionalization and therapeutic uses based on the target

Nanoparticles	Functionalization	Drug	Use	Refs.
Human serum albumin	Amino/acid group	Doxorubicin	Antineoplastic	[260]
Trimyristin	Sterically stabilized	Paclitaxel	Ovarian, lung, breast cancer	[261]
PLLA-b-PEG	Folate targeted	Doxorubicin	Solid tumors	[262]
PEG-PE	Lipid conjugated	Paclitaxel	Various cancers	[263]
PEG	Lipid conjugated	Tamoxifen	Lung carcinoma	[264]
Polymer-lipid hybrid	Lipid conjugated	Doxorubicin	Solid cancer	[265]
PCL-b-trimethylene carbonate-PEG	Serum protein	Ellipticin	Anticancer	[266]
PAMAM dendrimers	Folic acid	ethotrexate	Epithelial cancer	[267]
PEG	Albumin bound	Doxorubicin	Various cancers	[268]
Micelles	Biotin-antibody-conjugated	Daunomycin	Brain tumor	[269]
PLGA	Alendronate	Estrogen	Bone-osteoporosis	[270]
Poly(DEAP-Lys)-b-PEG -b-PLLA	Poly(lysine)	Doxorubicin	pH sensitive tumor	[271]
PLGA-b-PEG-COOH	PSMA	Anti cancer	Prostate- cancer	[272]
PEG or PE particles	Transferrin	Oligonucleotide	Brain- gene	[273]
PLLA-PEG	Biotin	Anti cancer	Cancer therapy	[274]
Polystyrol	Sc-TNF	Anti cancer	Cancer therapy	[275]
PLA	Aptamer	Anti cancer	Prostate cancer	[276]
PE	RGD peptides	siRNA	Vasculature cancer	[277]
mPEG/PLGA	Peptidomimetics	Anti cancer	Brain cells cancer	[278]
PLA	Galactose	Retinoic acid	Hepatocytes	[279]
PLGA	MP lipid A	Anti cancer	Dentritic cells	[280]

In particular, the non-invasive intranasal delivery offers large interests in the targeted route of administration. Nasal delivery helps drugs to bypass the blood-brain barrier and hence acts as an excellent platform for brain targeting. The intranasal drug delivery several approaches should be considered, attending, specifically, to the nature of pathological condition (acute or chronic) and intended effects of drug treatment (local, systemic or at CNS). Local delivery, nasal vaccines, systemic delivery and CNS delivery through nasal route is the prime route for drug administration to treat the various diseases. So the nasal vaccination is a promising alternative to the classic parenteral route because the nasal mucosa possesses abundant nasal associated lymphoid tissue (NALT), dentritic cells, large surface area, and low proteolytic enzymes that serve as a primary defense system against pathogens. It can exhibit high drug concentration, permeation, no first-pass effect and compliance administration without enzymatic destruction. Moreover, antigens encapsulated nanoparticles ensure enhanced uptake and controlled release of antigens from the nasal vasculature membrane with strong immunogenicity and improved systemic therapeutic responses. Also, the bio-nanotechnology applied to the parenteral administrations techniques such as microneedles, jet-injections, ultrasound, iontophoresis, and electrophoresis. Theses systems extend painless, patient-friendly alternatives to injections for the delivery of molecule [313-317]. Drug administration using microneedles for the transdermal delivery routes have been reported elsewhere [318-320]. Microneedles are arrays of micrometer-sized shallow needles that penetrate only into the superficial layers of skin, thereby eliminating the pain associated with standard hypodermic needles [321]. Microneedles have been made from a variety of materials and in particular the polymers have been shown to be effective. They have also been produced in solid and as well as in hollow forms. Solid microneedles are used to render skin permeable, whereas hollow microneedles actively deliver drugs into the skin at a controlled rate. In contrast, jet injectors deliver a high-velocity liquid jet stream into the skin, delivering drugs into various skin layers, depending on the jet parameters [322]. Jet injectors have a long history, particularly in the delivery of vaccines, insulin, and growth hormone. Ultrasound enhances skin permeability by cavitation, which temporarily disrupts skin structure [323]. Iontophoresis and electroporation use electric fields to alter the skin structure and/or provide additional driving force for drug penetration through the skin [324]. These new routes of administration of therapeutics with improved responses have been achieved by high drug concentration in target, permeation, no first-pass effect, high bioavailability and compliance administration without enzymatic destruction [325, 326].

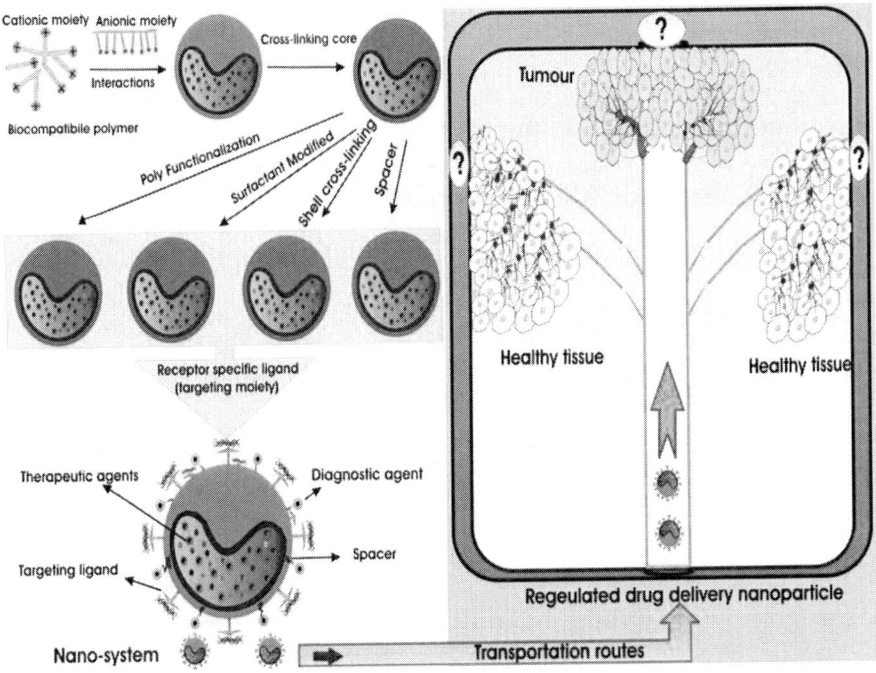

Figure 8. Schematic diagrams represent the various functionalizations of the nano-engines for smart drug delivery systems, which the pre-regulated nanoparticle recognizes the tumor cells not the healthy.

8. CONCLUSION

The uses of bio-nanotechnology in therapeutics a number of unexpected inventions have been done recently on polymer based nanometers, which have great attention in the field of smart drug delivery applications. The biomaterials including protein based polymers, polysaccharide based polymers, natural or synthetic or semi-synthetic polymers, various biomaterials and combination of polymer have utilized to prepare various kinds of nano-formulations towards the smart drug delivery applications. Several polymeric nanoparticle-based therapeutic systems have been established for the treatment of various diseases. Several nanoparticle based drug delivery systems have been approved in clinical trials, some of them in under pre-clinical trial levels, this nanoparticle based system can provide the increased half-life, high biocompatibility, and minimum immunogenicity, site targeting and overcome the membrane barriers. Also the last era, major and new identifications have been drastically established in the smart material that alter its own structure and function in response to the environment. This performance has been used for the fabrication smart drug delivery systems, Smart polymer matrices release drugs by environment responses this system have been successfully achieved. In parallel the new method of bottom-up and top-down nanofabrication technologies provided precisely controlled size and shaped nano-particulate delivery system. Simultaneously, various advanced significant routes of targeting have developed and successfully achieved to the site of action. At present, the field of microfluidics for synthesis, micro-needle for transdermal and site targeted delivery is still in its infancy. So the pharmaceutical industry has to bring these products into industry-led investigation and the improvement in this would possibly to quicken their progress.

9. FUTURE PERSPECTIVES

Although there are considerable amount researches have been done in the field of drug delivery so far. In the polymeric nanoparticle based drug therapy has to be enhanced by incorporating by the combination therapies, Smart delivery has been achieved successfully in the case of cancer, but need to be concentrating more on other pathologies, also numerous challenges remain. From the material viewpoint, most of the smart delivery systems mechanism do well in vitro studies but flops the in vivo studies. So the research has to be re-considering to come up with simple, straightforward, efficient and reasonably accurate preparations with broadly applicable strategies, the pharmacologically active agent targeting to pathological sites, for the development of smart drug delivery systems. In technology vice the research has to focus into the fusion technologies. Although several specific specialized technologies have been shown to in polymer synthesis, functionalization, analysis, in vitro and in vivo study in the field of polymer science, the combinations of two or more techniques are often more effective than single technologies like a combination of controlled radical polymerization with click chemistry. The fusion technologies can fulfil the various existing drawbacks of some individual technologies, and this has the high potentiality, synergistic enhancement in safest nanoparticle based drug delivery. Consider merging and adopting two or more right technologies for getting a high-throughput technology by selecting the right combinations is a fruitful area for research that is still largely unexplored. This new understanding must be incorporated into the future of newer polymeric based nanoparticle synthesis development and evaluation of smart drug delivery. Also the next generation of polymeric nanoparticle based delivery systems with drugs like growth factors, hormones, antibodies, genes, peptides, etc.; should also enhance the efficiency and minimize the unwanted effects.

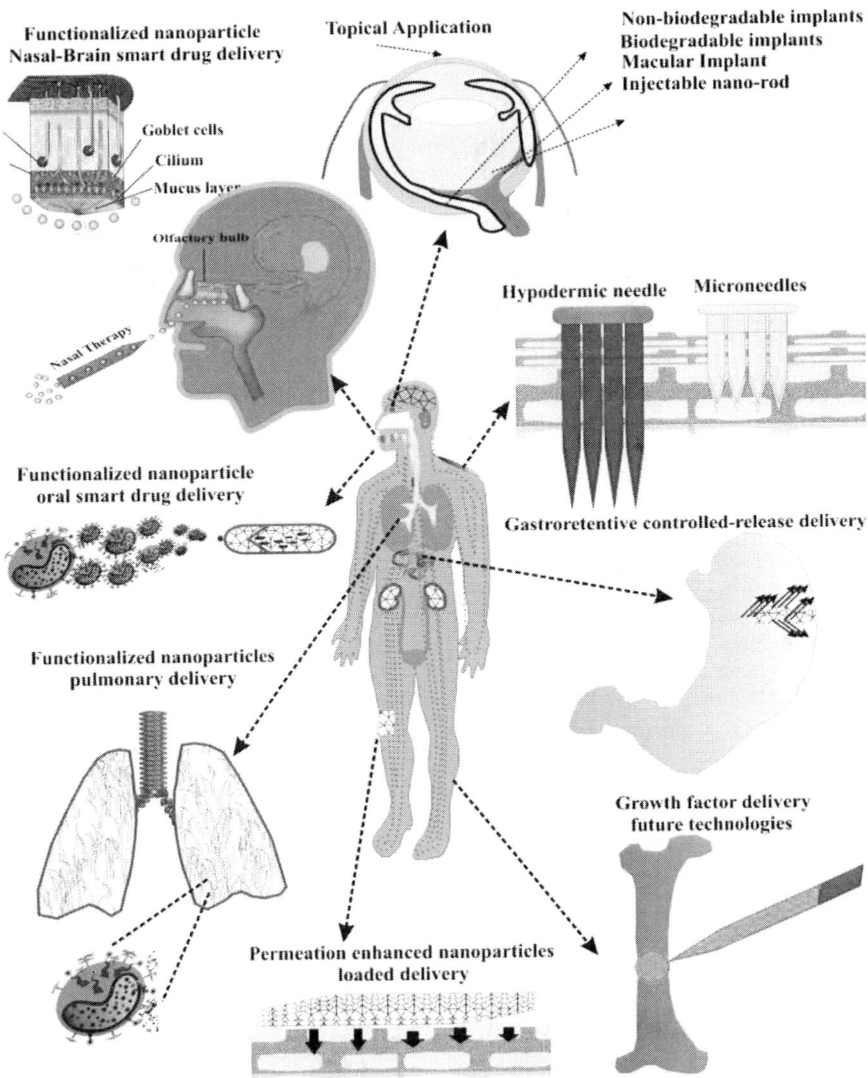

Figure 9. Schematic diagrams represent the recent developments of various significance routes of administration and targeting strategies

ACKNOWLEDGEMENTS

This work was supported by the R&D Program for Society of the National Research Foundation(NRF) funded by the Ministry of Science, ICT & Future Planning (2013M3C8A3078806 and 2013M3C1A8A01072922).

REFERENCES

1. Allemann E, Gurny R, Doelker E. Drug-Loaded Nanoparticles: Preparation Methods and Drug Targeting Issues. European Journal of Pharmaceutics and Biopharmaceutics 1993;39(5) 173-191.

2. Kawashima Y. Nanoparticulate Systems for Improved Drug Delivery. Advanced Drug Delivery Reviews 2001;47(1) 1-2.

3. Soppimath KS, Aminabhavi TM, Kulkarni AR, Rudzinski WE. Biodegradable Polymeric Nanoparticles as Drug Delivery Devices. Journal of Controlled Release 2001;70(1-2) 1-20.

4. Panyam J, Labhasetwar V. Biodegradable Nanoparticle from Drug and Gene Delivery to Cells and Tissue. Advanced Drug Delivery Reviews 2003;55(3) 329-347.

5. Brigger I, Dubernet C, Couvreur P. Nanoparticles in Cancer Therapy and Diagnosis. Advanced Drug Delivery Reviews 2002;54(5) 631-651.

6. Cui Z, Mumper RJ. Plasmid DNA-Entrapped Nanoparticles Engineered from Microemulsion Precursors: In Vitro and In Vivo Evaluation. Bioconjugate Chemistry 2002;13(6) 1319-1327.

7. Cohen H, Levy RJ, Gao J, Fishbein I, Kousaev V, Sosnowski S, Slomkowski S, Golomb G. Sustained Delivery and Expression of DNA Encapsulated in Polymeric Nanoparticles. Gene Therapy 2000;(22) 71896-1905.

8. Cao X, Lai S, Lee LJ. Design of a Self-Regulated Drug Delivery Device. Biomedical Micro devices 2001;3(2)109-118.

9. Kasagana VN, Karumuri SS, Thirumal. M. Recent Advances in Smart Drug Delivery Systems. International Journal of Pharmacy and Biotechnology 2011;1(3) 201-207.

10. Kreuter, J. Nanoparticles and Nanocapsules-New Dosage Forms in the Nanometer Range. Pharmaceutica Acta Helvetiae 1978;53(2) 33-39.

11. Kreuter J. Nanoparticles. In: Swarbrick J, Boylan JC. (ed.) Encyclopedia of Pharmaceutical Technology. Marcel Dekker Inc: New York, NY, 1994. p165-190.

12. Birrenbach G, Speiser PP. Polymerized Micelles and their use as Adjuvants in Immunology. Journal of Pharmaceutical Sciences 1976;65(12) 1763-1766.

13. Fresta M, Cavallaro G, Giammona G, Wehrli E, Puglisi G. Preparation and Characterization of Polyethyl-2-Cyanoacrylate Nanocapsules containing Antiepileptic Drugs. Biomaterials 1996;17(8) 751-758.

14. Vauthier-Holtzscherer C, Benabbou S, Spenlehauer G, Veillard M, Couvreur P. Methodology for the Preparation of Ultra-Dispersed Polymer Systems. STP Pharma Sciences 1991;1 109-116.

15. Al Khouri Fallouh N, Roblot-Treupel L, Fessi H, Devissaguet JP, Puisieux F. Development of a New Process for the Manufacture of Polyisobutylcyanoacrylate Nanocapsules. International Journal of Pharmaceutics 1986;28(2-3) 125-132.

16. Rollot JM, Couvreur P, Roblot-Treubel L, Puisieux F. Physicochemical and Morphological Characterization of Polyisobutyl Cyanoacrylate Nanocapsules. Journal of Pharmaceutical Sciences 1986;75(4) 361-364.

17. Legrand P, Barratt G, Mosqueira V, Fessi H, Devissaguet JP. Polymeric Nanocapsules as Drug Delivery Systems. A review. STP Pharma Sciences 1999;9(5) 411-418.

18. Barratt GM. Therapeutic Applications of Colloidal Drug Carriers. Pharmaceutical Science and Technology Today 2000;3(5) 163-171.

19. Fresta M, Cavallaro G, Giammona G, Wehrli E, Puglisi G. Preparation and Characterization of Polyethyl-2-Cyanoacrylate Nanocapsules containing Antiepileptic Drugs. Biomaterials 1996;17(8) 751-758.

20. Schroeder U, Sommerfeld P, Ulrich S, Sabel BA. Nanoparticle Technology for Delivery of Drugs across the Blood-Brain Barrier. Journal of Pharmaceutical Sciences 1998;87(11) 1305-1307

21. Hans ML. Lowman AM. Biodegradable Nanoparticles for Drug Delivery and Targeting. Current Opinion in Solid State and Materials Science 2002;6(4) 319-327.

22. Arshady R. Preparation of Biodegradable Microspheres and Microcapsules: 2. Polylactides and Related Polyesters. Journal of Controlled Release 1991;17(1) 1-22.

23. Bodmeier R, Chen H, Tyle P, Jarosz P. Spontaneous Formation of Drug-containing Acrylic Nanoparticles. Journal of Microencapsulation 1991;8(2) 161-170.

24. Niwa T, Takeuchi H, Hino T, Kunou N, Kawashima Y. Preparations of Biodegradable Nanospheres of Water-Soluble and Insoluble Drugs with D,L-Lactide/Glycolide Copolymer by a Novel Spontaneous Emulsification Solvent Diffusion Method, and the Drug Release Behaviour. Journal of Controlled Release 1993;25(1-2) 89-98.

25. Leroux J-C, Allemann E, Doelker E, Gurny R. New Approach for the Preparation of Nanoparticles by an Emulsification-Diffusion Method. European Journal of Pharmaceutics and Biopharmaceutics 1995;41(3) 14-18.

26. Quintanar-Guerrero D, Fessi H, Allemann E, Doelker E. Influence of Stabilizing Agents and Preparative Variables on the Formation of Poly(D,L-Lactic Acid) Nanoparticles by an Emulsification-Diffusion Technique. International Journal of Pharmaceutics 1996;143(2) 133-141.

27. Kawashima Y, Niwa T, Handa T, Takeuchi H, Iwamoto T, Itoh K. Preparation Of Controlled-Release Microspheres of Ibuprofen with Acrylic Polymers by a Novel Quasi-Emulsion Solvent Diffusion Method. Journal of Pharmaceutical Sciences 1989;78(7) 68-72.

28. Kawashima Y, Iwamoto T, Niwa T, Takeuchi H, Hino T. Size Control of Ibuprofen Microspheres with an Acrylic Polymer by Changing the pH In an Aqueous Dispersion Medium and its Mechanism. Chemical and Pharmaceutical Bulletin 1993;41(1) 191-195.

29. Quintanar-Guerrero D, Allemann E, Doelker E, Fessi H. Preparation and Characterization of Nanocapsules from Preformed Polymers by a New Process based on Emulsification-Diffusion Technique. Pharmaceutical Research 1998;15(7) 1056-1062.

30. Allemann E, Doelker E, Gurny R. Drug Loaded Poly(lactic acid) Nanoparticles Produced by a Reversible Salting-out Process: Purification of an Injectable Dosage Form. European Journal of Pharmaceutics and Biopharmaceutics 1993;39 13-18.

31. Allemann E, Gurny R, Doelker E. Preparation of Aqueous Polymeric Nanodispersions by a Reversible Salting-Out Process: Influence of Process Parameters on Particle Size. International Journal of Pharmaceutics 1992;87(1-3) 247-253.

32. Rodrigues Jr JM, Fessi H, Bories C, Puisieux F, Devissaguet JP. Primaquine-Loaded Poly(Lactide) Nanoparticles: Physicochemical Study and Acute Tolerance in Mice. International Journal of Pharmaceutics 1995;126(1-2) 253-60.

33. Fessi H, Puisieux F, Devissaguet JP, Ammoury N, Benita S. Nanocapsule Formation by Interfacial Polymer Deposition Following Solvent Displacement. International Journal of Pharmaceutics 1989;55(1) R1-R4

34. Peltonen L, Koistinen P, Karjalainen M, Hakkinen A, Hirvonen J. The Effect of Cosolvents on the Formulation of Nanoparticles from Low-Molecular-Weight Poly(L)Lactide. AAPS PharmSciTech 2002;3(4) E32.

35. Niwa T, Takeuchi H, Hino T, Nohara M, Kawashima Y. Biodegradable Submicron Carriers for Peptide Drugs: Preparation of Dl-Lactide/Glycolide Copolymer (PLGA) Nanospheres with Nafarelin Acetate by a Novel Emulsion-Phase Separation Method in an Oil System. International Journal of Pharmaceutics 1995;121(1) 45-54.

36. Chen X, Young TJ, Sarkari M, Williams RO 3rd, Johnston KP. Preparation of Cyclosporine a Nanoparticle by Evaporative Precipitation into Aqueous Solution. International Journal of Pharmaceutics 2002;242(1-2) 3-14.

37. Sarkari M, Brown J, Chen X, Swinnea S, Williams RO 3rd, Johnston KP. Enhanced Drug Dissolution using Evaporative Precipitation into Aqueous Solution. International Journal of Pharmaceutics 2002;243(1-2) 17-31.

38. Gomez A, Bingham D, de Juan L, Tang K. Production of Protein Nanoparticles by Electrospray Drying. Journal of Aerosol Science 1998;29(5-6) 561-574.

39. Liversidge GG, Cundy KC. Particle Size Reduction for Improvement of Oral Bioavailability of Hydrophobic Drugs: I. Absolute Oral Bioavailability of Nanocrystalline Danazol in Beagle Dogs. International Journal of Pharmaceutics 1995;125(1) 91-97.

40. Jacobs C, Kayser O, Muller RH. Nanosuspensions as a New Approach for the Formulation for the Poorly Soluble Drug Tarazepide. International Journal of Pharmaceutics 2000;196(2) 161-164.

41. Merisko-Liversidge E, Sarpotdar P, Bruno J, Hajj S, Wei L, Peltier N, Rake J, Shaw JM, Pugh S, Polin L, Jones J, Corbett T, Cooper E, Liversidge GG. Formulation and Antitumor Activity Evaluation of Nanocrystalline Suspensions of Poorly Soluble Anticancer Drugs. Pharmaceutical Research 1996;13(2) 272-278.

42. Jacobs C, Kayser O, Muller RH. Production and Characterisation of Mucoadhesive Nanosuspensions for the Formulation of Bupravaquone. International Journal of Pharmaceutics 2001;214(1-2) 3-7.

43. Merisko-Liversidge E, Liversidge GG, Cooper ER. Nanosizing: A Formulation Approach for Poorly-Water-Soluble Compounds. European Journal of Pharmaceutical Sciences 2003;18(2) 113-120.

44. Muller CR, Schaffazick SR, Pohlmann AR, de Lucca Freitas L, Pesce da Silveira N, Dalla Costa T, Guterres SS. Spray-Dried Diclofenac-Loaded Poly(E-Caprolactone) Nanocapsules and Nanospheres. Preparation and Physicochemical Characterization. Pharmazie 2001;56(11) 864-867.

45. Vergote GJ, Vervaet C, Van Driessche I, Hoste S, De Smedt S, Demeester J, Jain RA, Ruddy S, Remon JP. An Oral Controlled Release Matrix Pellet Formulation Containing Nanocrystalline Ketoprofen. International Journal of Pharmaceutics 2001;219(1-2) 81-87.

46. Bodmeier R, Wang H, Dixon DJ, Mawson S, Johnston KP. Polymeric Microspheres Prepared by Spraying into Compressed Carbon Dioxide. Pharmaceutical Research 1995;12(8) 1211-1217.

47. Reverchon E. Supercritical Antisolvent Precipitation of Micro-and Nano-Particles. Journal of Supercritical Fluids 1999;15(1) 1-21.

48. Reverchon E. Della Porta G. Production of Antibiotic Micro-and Nano-Particles by Supercritical Antisolvent Precipitation. Powder Technology 1999;106 23-29.

49. Reverchon E, Marco De I, Della Porta G. Rifampicin Micoparticles Production by Supercritical Antisolvent Precipitation. International Journal of Pharmaceutics 2002;243(1-2) 83-91.

50. Rogers TL, Johnston KP, Williams RO 3rd. Solution-based Particle Formation of Pharmaceutical Powders by Supercritical or Compressed

Fluid Co2 and Cryogenic Spray-Freezing Technologies. Drug Development and Industrial Pharmacy 2001;27(10) 1003-1015.

51. Tom JW, Debenedetti PG. Particle Formation with Supercritical Fluids-A Review. Journal of Aerosol Science 1991;22(5) 555-584.

52. Tu LS, Dehghani F, Forster NR. Micronisation and Microencapsulation of Pharmaceuticals using a Carbon Dioxide Antisolvent. Powder Technology 2002;126(2) 134-149

53. Broadhead J, Rouan SKE, Rhodes CT. The Spray Drying of Pharmaceuticals. Drug Development and Industrial Pharmacy 1992;18(11-12) 1169-1206.

54. Giunchedi P, Conte U. Spray-Drying as a Preparation Method of Microparticulate Drug Delivery System: An Overview. STP Pharma Sciences 1995;5 276-290.

55. Nielsen F. Spray Drying Pharmaceuticals. Manufacturing Chemist 1982;57 38-41.

56. Esposito E, Cervellati F, Menegatti E, Nastruzzi C, Cortesi R. Spray Dried Eudragit Microparticles as Encapsulation Devices for Vitamin C. International Journal of Pharmaceutics 2002;242(1-2) 329-334.

57. Bodmeier, R. and Chen, H. Preparation of Biodegradable Poly(+)Lactide Microparticles using a Spray-Drying Technique. Journal of Pharmacy and Pharmacology 1988;40(11) 754-757.

58. Wang FJ, Wang CH. Sustained Release of Etanidazole from Spray Dried Microspheres Prepared by Non-Halogenated Solvents. Journal of Controlled Release 2002;81(3) 263-280.

59. Pignatello R, Vandelli MA, Giunchedi P, Puglisi G. Properties of Tolmetin-Loaded Eudragit RL100 and Eudragit RS100 Microparticles Prepared by Different Techniques. STP Pharma Sciences 1997;7 148-157.

60. Magenheim B, Benita S. Nanoparticle Characterization: A Comprehensive Physicochemical Approach. STP Pharma Sciences 1991;1 221-241.

61. Muller RH, Jacobs C, Kayser O. Nanosuspensions as Particulate Drug Formulations in Therapy Rationale for Development and what can we expect for the Future. Advanced Drug Delivery Reviews 2001;47(1) 3-19.

62. Palakodaty S, York P. Phase Behavioural Effects on Particle Formation Processes using Supercritical Fluids. Pharmaceutical Research 1999;16(7) 976-985.

63. Kong DF, Goldschmidt-Clermont PJ. Tiny Solutions for Giant Cardiac Problems. Trends in Cardiovascular Medicine 2005;15(6)207-11.

64. Khafagy el S, Morishita M, Onuki Y, Takayama K. Current Challenges in Non-Invasive Insulin Delivery Systems: A Comparative Review. Advanced Drug Delivery Reviews 2007;59(15) 1521-46.

65. Ferrari M. Cancer Nanotechnology: Opportunities and Challenges. Nature Reviews. Cancer 2005;5(3) 161-71.

66. Duncan R. Polymer Conjugates as Anticancer Nanomedicines. Nature Reviews. Cancer 2006; 6(9) 688-701.

67. Nasir A. Nanodermatology: A Bright Glimpse just Beyond the Horizon-Part I. Skin Therapy Letter 2010;15(8) 1-4.

68. Gupta S, Bansal R, Gupta S, Jindal N, Jindal A. Nanocarriers and Nanoparticles for Skin Care and Dermatological Treatments. Indian Dermatology Online Journal 2013;4(4) 267-272.

69. Silva GA. Neuroscience Nanotechnology: Progress, Opportunities and Challenges. Nature Reviews. Neuroscience 2006;7(1) 65-74.

70. Pardridge WM, Blood-Brain Barrier Delivery. Drug Discovery Today 2007;12(1-2) 54-61.

71. Nuxoll E. BioMEMS in Drug Delivery. Advanced Drug Delivery Reviews 2013;65(11-12) 1611-25.

72. Wang Y, Byrne JD., Napier ME, DeSimone JM. Engineering Nanomedicines using Stimuli-Responsive Biomaterials. Advanced Drug Delivery Reviews 2012;64(11) 1021-1030.

73. Shim TS, Kim S, Yang S, Elaborate Design Strategies Toward Novel Microcarriers for Controlled Encapsulation and Release. Particle and Particle Systems Characterization 2013;30(1) 9-45.

74. Christopher GF, Anna SL. Microfluidic Methods for Generating Continuous Droplet Streams. Journal of Physics D: Applied Physics 2007;40(19) R319-R336.

75. Dendukuri D, Doyle PS. The Synthesis and Assembly of Polymeric Microparticles using Microfluidics. Advanced Materials 2009;21(41) 4071-4086.

76. Valencia PM, Farokhzad OC, Karnik R, Langer R. Microfluidic Technologies for Accelerating the Clinical Translation of Nanoparticles. Nature Nanotechnology 2012;7(10) 623-9.

77. Choi DH, Subbiah R, Kim IH, Han DK, Park K. Dual Growth Factor Delivery using Biocompatible Core-Shell Microcapsules for Angiogenesis. Small 2013;9(20) 3468-76.

78. Luttge R, Berenschot E, De Boer M, Altpeter D, Vrouwe E, Van den Berg A, Elwenspoek M. Integrated Lithographic Molding for Microneedle-Based Devices. Journal of Microelectromechanical Systems 2007;6(4) 872-884.

79. Park J, Allen M, Prausnitz M, Biodegradable Polymer Microneedles: Fabrication, Mechanics and Transdermal Drug Delivery, Journal of Controlled Release 2005;104(1) 51-66.

80. Wendorf JR, Ghartey-Tagoe EB, Williams SC, Enioutina E, Singh P, Cleary GW, Transdermal Delivery of Macromolecules using Solid-State Biodegradable Microstructures. Pharmaceutical Research 2011;28(1) 22-30.

81. Raphael AP, Prow TW, Crichton ML, Chen X, Fernando GJP, Kendall MAF, Targeted Needle-Free Vaccinations in Skin Using Multilayered, Densely-Packed Dissolving Microprojection Arrays. Small 2010;6(16) 1785-1793.

82. Kronenberg HM. Developmental Regulation of the Growth Plate. Nature 2003;423(6937) 332-336.

83. Karsenty G. The Complexities of Skeletal Biology. Nature 2003;423(6937) 316-318.

84. Zhuang J, Wu H, Yang Y, Cao YC. Controlling Colloidal Superparticle Growth through Solvophobic Interactions. Angewandte Chemie 2008;47(12) 2208-2212.

85. Manoharan VN, Elsesser MT, Pine DJ. Dense Packing and Symmetry in Small Clusters of Microspheres. Science 2003;301(5632) 483-487.

86. Brannon-Peppas L. Recent Advances on the use of Biodegradable Microparticles Andnanoparticles in Controlled Drug Delivery. International Journal of Pharmaceutics 1995;116(1) 1-9.

87. Marimuthu M, Bennet D, Kim S. Self-assembled Nanoparticles OF PLGA-Conjugated Glucosamine as a Sustained Transdermal Drug Delivery Vehicle. Polymer Journal 2013;45 202-209

88. Kopecek J, Kopeckova P, Minko T, Lu ZR, Peterson CM. Water Soluble Polymers in Tumor Targeted Delivery. Journal of Controlled Release 2001;74(1-3) 147-158.

89. Nakanishi T, Fukushima S, Okamoto K, Suzuki M, Matsumura Y, Yokoyama M, Okano T, Sakurai Y, Kataoka K. Development of the Polymer Micelle Carrier System for Doxorubicin. Journal of Controlled Release 2001;74(1-3) 295-302,

90. Tian Y, Bromberg L, Lin SN, Alan Hatton T, Tam KC. Complexation and Release of Doxorubicin from its Complexes with Pluronic P85-b-Poly(Acrylic Acid) Block Copolymers. Journal of Controlled Release 2007;121(3) 137-145.

91. Cao T, Munk P, Ramireddy C, Tuzar Z, Webber SE. Fluorescence Studies of Amphiphilic Poly(Methacrylic Acid)-Block-Polystyrene-Block-Poly(Methacrylic Acid) Micelles. Macromolecules 1991;24(23) 6300-6305.

92. Zhang L, Yu K, Eisenberg A. Ion-Induced Morphological Changes in "Crew-Cut" Aggregates of Amphiphilic Block Copolymers. Science 1996;272(5269) 1777-1779.

93. Discher DE, Eisenberg A. Polymer Vesicles. Science 2002;2979(5583) 967-973.

94. Lee CM, Tanaka T, Murai T, Kondo M, Kimura J, Su W, Kitagawa T, Ito T, Matsuda H, Miyasaka M. Novel Chondroitin Sulfate-Binding Cationic Liposomes Loaded with Cisplatin Efficiently Suppress the Local Growth and Liver Metastasis of Tumor Cells In Vivo. Cancer Research 2002;62(15) 4282-4288.

95. Gabizon AA. Selective Tumor Localization and Improved Therapeutic Index of Anthracyclines Encapsulated in Long-Circulating Liposomes. Cancer Research 1992;52(4) 891-896.

96. Cleland JL, Lim A, Barron L, Duenas ET, Powell MF. Development of a Single-Shot Subunit Vaccine for HIV-1: Part 4. Optimizing

Microencapsulation and Pulsatile Release of MN rgp120 from Biodegradable Microspheres. Journal of Controlled Release 1997;47(2) 135-150.

97. Rilling P, Walter T, Pommershein R, Vogt W. Encapsulation of Cytochrome C by Multilayer Microcapsules. A Model for Improved Enzyme Immobilization. Journal of Membrane Science 1997;129(2) 283-287.

98. Fonseca T, Relogio P, Martinho JMG, Farinha JPS. Preparation and Surface Characterization of Polymer Nanoparticles Designed for Incorporation into Hybrid Materials. Langmuir 2007;23(10) 5727-5734.

99. Sajjadi S. Population Balance Modeling of Particle Size Distribution in Monomer-Starved Semibatch Emulsion Polymerization. AIChE Journal 2009;55(12) 3191-3205.

100. Hui Wang, Garry L, Rempel. pH-Responsive Polymer Core-Shell Nanospheres for Drug Delivery. Journal of Polymer Science, Part A: Polymer Chemistry 2013;51(20) 4440-4450

101. Valencia PM, Farokhzad OC, Karnik R, Langer R. Microfluidic Technologies for Accelerating the Clinical Translation of Nanoparticles. Nature Nanotechnology 2012;7(10) 623-9.

102. DeMello AJ. Control and Detection of Chemical Reactions in Microfluidic Systems. Nature 2006;442(7101) 394-402.

103. Capretto L, Cheng W, Hill M, Zhang X. Micromixing within Microfluidic Devices. Topics in Current Chemistry 2011;304 27-68.

104. Rhee M, Valencia PM, Rodriguez MI, Langer R, Farokhzad OC, Karnik R. Synthesis of Size-Tunable Polymeric Nanoparticles Enabled by 3D Hydrodynamic Flow Focusing in Single-Layer Microchannels. Advanced Materials 2011;23(12) H79–H83.

105. Liu K, Wang H, Chen KJ, Guo F, Lin WY, Chen YC, Phung DL, Tseng HR, Shen CK. A Digital Microfluidic Droplet Generator Produces Self-Assembled Supramolecular Nanoparticles for Targeted Cell Imaging. Nanotechnology 2010;21(44) 445603.

106. Valencia PM, Basto PA, Zhang L, Rhee M, Langer R, Farokhzad OC, Karnik R. Single-Step Assembly of Homogenous Lipid-Polymeric and Lipid-Quantum Dot Nanoparticles Enabled by Microfluidic Rapid Mixing. ACS Nano 2010;4(3) 1671-1679.

107. Jahn A, Reiner JE, Vreeland WN, DeVoe DL, Locascio LE, Gaitan M. Preparation of Nanoparticles by Continuous-Flow Microfluidics. Journal of Nanoparticle Research 2008;10(6) 925-934.

108. Bennet D, Kim S. A Transdermal Delivery System to Enhance Quercetin Nanoparticle Permeability, Journal of Biomaterials Science. Polymer Edition 2013;24(2) 185-209.

109. Bennet D, Marimuthu M, Kim S, An J. Dual Drug-Loaded Nanoparticles on Self-Integrated Scaffold for Controlled Delivery. International Journal of Nanomedicine 2012;2012(7) 3399-419.

110. Srinivasa PC, Ramesh MN, Tharanathan RN. Effect of Plasticizers and Fatty Acids on Mechanical and Permeability Characteristics of Chitosan Films. Food Hydrocolloids 2007;21(7) 1113-22.

111. Zhang ML, Gong XH, Zhao YD, Zhang NM. Properties and Biocompatibility of Chitosan Films Modified by Blending with PEG. Biomaterials 2002;23(13) 2641-8

112. Badet C, Furiga A, Thebaud N. Effect of Xylitol on an In Vitro Model of Oral Biofilm. Oral Health & Preventive Dentistry 2008;6(4) 337-41.

113. Chandy T, Mooradian DL, Rao GH. Chitosan/Polyethylene Glycol-Alginate Microcapsules for Oral Delivery of Hirudin. Journal of Applied Polymer Science 1998;70(11) 2143-2153.

114. Rajangam T, Paik HJ, An SSA. Development of Fibrinogen Microspheres as a Biodegradable Carrier for Tissue Engineering. BioChip Journal 2011;5(2) 175-183.

115. Bromberg LE, Ron ES. Temperature-Responsive Gels and Thermogelling Polymer Matrices for Protein and Peptide Delivery. Advanced Drug Delivery Reviews 1998;31(3)197-221.

116. Jain GK, Sharma AK, Agrawal SS. Transdermal Controlled Administration of Verapamil-Enhancement of Skin Permeability. International Journal of Pharmaceutics 1996;130(2) 169-177.

117. Ding H, Triggle CR. Endothelial Cell Dysfunction and the Vascular Complications Associated with Type 2 Diabetes: Assessing the Health of the Endothelium. Vascular Health and Risk Management 2005;1(1) 55-71.

118. Chien YW. Novel Drug Delivery Systems: 2nd ed, Revised and Expanded; Marcel Dekker, Inc.: New York, 1992.

119. Chien YW. Rate-Control Drug Delivery Systems: Controlled Release vs. Sustained Release. Medical Progress Through Technology 1989;15(1-2) 21-46.

120. Robinson JR. Sustained and Controlled Release Drug Delivery Systems; Marcel Dekker, Inc.: New York, 1978.

121. Hsieh DST. Subcutaneous Controlled Delivery of Estradiol by Compudose Implants: In Vitro and In Vivo Evaluations. Drug Development and Industrial Pharmacy 1987;13(15) 2651-2666.

122. Segal SJ. The Development of NORPLANT Implants. Studies in Family Planning 1983;14(6-7) 159-163.

123. Diaz S, Pavez M, Miranda P, Robertson DN, Sivin I, Croxatto HB. A five-Year Clinical Trial of Levonorgestrel Silastic Implants (Norplant 2). Contraception 1982;25(5) 447-456.

124. Weiner E, Victor A, Johansson ED. Plasma Levels of D-Norgestrel after Oral Administration. Contraception 1976;14(5) 563-570.

125. Croxatto HB, Diaz S, Miranda P, Elamsson K, Johansson ED. Plasma Levels of Levonorgestrel in Women during Longterm use of Norplant. Contraception 1981;23(2) 197-209.

126. Chien YW. Microsealed Drug Delivery Systems: Theoretical Aspects and Biomedical Assessments. In Recent Advances in Drug Delivery Systems; Anderson, JM, Kim, SW. Ed. Plenum Press: New York; 1984. p367-387.

127. Sunil K, Anil K, Vaibhav G, Kuldeep M, Pankaj R. Oral Extended Release Drug Delivery System A Promising Approach. Asian Pharma Press. Asian Journal of Research in Pharmaceutical Sciences 2012;2(3) 101-106.

128. Chien YW. Novel Drug Delivery Systems: Fundamental, Developmental Concepts and Biomedical Assessments; Marcel Dekker, Inc.: New York, 1982.

129. Haznar-Garbacz D, Garbacz G, Eisenacher F, Klein S, Weitschies W. A Novel Liquefied Gas based Oral Controlled Release Drug Delivery System for Liquid Drug Formulations. European journal of Pharmaceutics and Biopharmaceutics 2012;81(2) 334-8.

130. Rajesh A, Jaimin P, Sangeeta A. A Novel Approach: Pulsatile Drug Delivery System. International Research Journal of Pharm 2012;3(9) 43-49

131. Hsieh, DST, Langer R. Zero-order Drug Delivery Systems with Magnetic Control. In Controlled Release Delivery Systems; Roseman TJ, Mansdorf SZ. Ed. Marcel Dekker, Inc.: New York, 1983.

132. Tyle P, Agrawala P. Drug Delivery by Phonophoresis. Pharmaceutical Research 1989;6(5) 355-361.

133. Dhote V, Bhatnagar P, Mishra PK, Mahajan SC, Mishra DK. Iontophoresis: A Potential Emergence of a Transdermal Drug Delivery System. Scientia pharmaceutica 2012;80(1) 1-28.

134. Okahata Y, Hachiya S, Ariga K, Seki T. The Electrical Breakdown and Permeability Control of a Bilayer-Corked Capsule Membrane in an External Electric Field. Journal of the American Chemical Society 1986;108(11) 2863-2869.

135. Kwon IC, Bae YH, Kim SW. Electrically Erodible Polymer Gel for Controlled Release of Drugs. Nature 1991;354(6350) 291-293.

136. Foss AC, Goto T, Morishita M, Peppas NA. Development of Acrylic-based Copolymers for Oral Insulin Delivery. European Journal of Pharmaceutics and Biopharmaceutics 2004;57(2) 163-169.

137. Ramachandran S, Shaheedha SM, Thirumurugan G, Dhanaraju MD. Floating Controlled Drug Delivery System of Famotidine Loaded Hollow Microspheres (Microballoons) in the Stomach. Current Drug Delivery 2010;7(1) 93-7.

138. Chu LY, Yamaguchi T, Nakao S. A Molecular-Recognition Microcapsule for Environmental Stimuli-Responsive Controlled Release. Advanced Materials 2002;14(5) 386-389.

139. Heller J. Biodegradable Polymers in Controlled Drug Delivery. Critical Reviews in Therapeutic Drug Carrier Systems 1984;1(1) 39-90.

140. Morimoto Y, Fujimoto S. Albumin Microspheres as Drug Carriers. Critical Reviews in Therapeutic Drug Carrier Systems 1985;2(1) 19-63.

141. Veronese FM, Schiavon O, Pasut G, Mendichi R, Andersson L, Tsirk A, Ford J, Wu G, Kneller S, Davies J, Duncan R. PEG-Doxorubicin Conjugates: Influence of Polymer Structure on Drug Release, In Vitro

Cytotoxicity, Biodistribution, and Antitumor Activity. Bioconjugate Chemistry 2005;16(4) 775-784.

142. You L-C, Lu F-Z, Li Z-C, Zhang W, Li F-M, Glucose-Sensitive Aggregates Formed by Poly(Ethylene Oxide)-Block-Poly(2-Glucosyloxyethyl Acrylate) with Concanavalin A in Dilute Aqueous Medium. Macromolecules 2003;36(1) 1-4.

143. Prausnitz MR, Langer R. Transdermal Drug Delivery. Nature biotechnology 2008;26(11) 1261-8.

144. Kaneko Y, Nakamura S, Sakai K, Kikuchi A, Aoyagi T, Sakurai Y, Okano T. Synthesis and Swelling-Deswelling Kinetics of Poly(n-Isopropylacrylamide) Hydrogels Grafted with LCST Modulated Polymers. Journal of Biomaterials Science. Polymer Edition 1999;10(11) 1079-1091.

145. Nakayama M, Okano T, Miyazaki T, Kohori F, Sakai K, Yokoyama M. Molecular Design of Biodegradable Polymeric Micelles for Temperature-Responsive Drug Release, Journal of Controlled Release 2006;115(1) 46-56.

146. Irie M. Stimuli-Responsive Poly(N-Isopropylacrylamide). Photo and Chemical-Induced Phase Transitions. Advances in Polymer Science 1993;110 49-65.

147. Suzuki A, Tanaka T. Phase Transition in Polymer Gels Induced by Visible Light. Nature 1990;346 345-347.

148. Thambi T, Deepagan VG, Yoon HY, Han HS, Kim SH, Son S, Jo DG, Ahn CH, Suh YD, Kim K, Kwon IC, Lee DS, Park JH. Hypoxia-Responsive Polymeric Nanoparticles for Tumor-Targeted Drug Delivery. Biomaterials 2014;35(5) 1735-43.

149. Schmaljohann D. Thermo-and pH-Responsive Polymers in Drug Delivery. Advanced Drug Delivery Reviews 2006;58(15)1655-70.

150. Wei H, Zhang XZ, Cheng H, Chen WQ, Cheng SX, Zhuo RX. Self-Assembled Thermo-and pH Responsive Micelles of Poly(10-Undecenoic Acid-B-N-Isopropylacrylamide) for Drug Delivery. Journal of Controlled Release 2006;116(3) 266-74.

151. Medeiros SF, Santos AM, Fessi H, Elaissari A. Stimuli-Responsive Magnetic Particles for Biomedical Applications. International Journal of Pharmaceutics 2011;403(1-2) 139-61.

152. Cheng R, Meng F, Deng C, Klok HA, Zhong Z. Dual and Multi-Stimuli Responsive Polymeric Nanoparticles for Programmed Site-Specific Drug Delivery. Biomaterials 2013;34(14) 3647-57.

153. Torchilin V. Tumor Delivery of Macromolecular Drugs based on the EPR Effect. Advanced Drug Delivery Reviews 2011;63(3) 131-5.

154. Hrkach J, Von Hoff D, Mukkaram Ali M, Andrianova E, Auer J, Campbell T, De Witt D, Figa M, Figueiredo M, Horhota A, Low S, McDonnell K, Peeke E, Retnarajan B, Sabnis A, Schnipper E, Song JJ, Song YH, Summa J, Tompsett D, Troiano G, Van Geen Hoven T, Wright J, LoRusso P, Kantoff PW, Bander NH, Sweeney C, Farokhzad OC, Langer R, Zale S. Preclinical Development and Clinical Translation of a PSMA-Targeted Docetaxel Nanoparticle with a Differentiated Pharmacological Profile. Science Translational Medicine 2012;4(128) 128ra139.

155. Gong J, Chen MW, Zheng Y, Wang SP, Wang YT. Polymeric Micelles Drug Delivery System in Oncology. Journal of Controlled Release 2012;159(3) 312-23.

156. Deng C, Jiang YJ, Cheng R, Meng FH, Zhong ZY. Biodegradable Polymeric Micelles for Targeted and Controlled Anticancer Drug Delivery: Promises, Progress and Prospects. Nano Today 2012;7(5) 467-80.

157. Meng FH, Cheng R, Deng C, Zhong ZY. Intracellular Drug Release Nanosystems. Materials Today 2012;15(10) 436-42.

158. Ganta S, Devalapally H, Shahiwala A, Amiji M. A Review of Stimuli-Responsive Nanocarriers for Drug and Gene Delivery. Journal of Controlled Release 2008;126(3) 187-204.

159. Meng FH, Zhong ZY, Feijen J. Stimuli-Responsive Polymersomes for Programmed Drug Delivery. Biomacromolecules 2009;10(2) 197-209.

160. Rapoport N. Physical Stimuli-Responsive Polymeric Micelles for Anti-Cancer Drug Delivery. Progress in Polymer Science 2007;32(8-9) 962-90.

161. Chen W, Meng FH, Li F, Ji SJ, Zhong ZY. pH-Responsive Biodegradable Micelles based on Acid-Labile Polycarbonate Hydrophobe: Synthesis and Triggered Drug Release. Biomacromolecules 2009;10(7) 1727-35.

162. Du JZ, Tang YQ, Lewis AL, Armes SP. pH-Sensitive Vesicles based on a Biocompatible Zwitterionic Diblock Copolymer. Journal of the American Chemical Society 2005;127(51) 17982-3.

163. Saito G, Swanson JA, Lee KD. Drug Delivery Strategy Utilizing Conjugation via Reversible Disulfide Linkages: Role and Site of Cellular Reducing Activities. Advanced Drug Delivery Reviews 2003;55(2) 199-215.

164. Meng FH, Hennink WE, Zhong ZY. Reduction-Sensitive Polymers and Bio-Conjugates for Biomedical Applications. Biomaterials 2009;30(12) 2180-98.

165. Cheng R, Feng F, Meng FH, Deng C, Feijen J, Zhong ZY. Glutathione-Responsive Nano-Vehicles as a Promising Platform for Targeted Intracellular Drug and Gene Delivery. Journal of Controlled Release 2011;152(1) 2-12.

166. Kuppusamy P, Li H, Ilangovan G, Cardounel AJ, Zweier JL, Yamada K, Krishna MC, Mitchell JB. Noninvasive Imaging of Tumor Redox Status and its Modification by Tissue Glutathione Levels. Cancer Research 2002;62(1) 307-12.

167. Chan A, Orme RP, Fricker RA, Roach P. Remote and Local Control of Stimuli Responsive Materials for Therapeutic Applications. Advanced Drug Delivery Reviews 2013;65(4) 497-514.

168. Alvarez-Lorenzo C, Bromberg L, Concheiro A. Light-Sensitive Intelligent Drug Delivery Systems. Photochemistry and Photobiology 2009;85(4) 848-60.

169. Chilkoti A, Dreher MR, Meyer DE, Raucher D. Targeted Drug Delivery by Thermally Responsive Polymers. Advanced Drug Delivery Reviews 2002;54(5) 613-30.

170. Wei H, Cheng SX, Zhang XZ, Zhuo RX. Thermo-Sensitive Polymeric Micelles based on Poly(N-Isopropylacrylamide) as Drug Carriers. Progress in Polymer Science 2009;34(9) 893-910.

171. Arruebo M, Fernandez-Pacheco R, Ibarra MR, Santamaria J. Magnetic Nano-Particles for Drug Delivery. Nano Today 2007;2(3) 22-32.

172. Sun C, Lee JSH, Zhang M. Magnetic Nanoparticles in MR Imaging and Drug Delivery. Advanced Drug Delivery Reviews 2008;60(11) 1252-65.

173. Xu HF, Meng FH, Zhong ZY. Reversibly Crosslinked Temperature-Responsive Nano-Sized Polymersomes: Synthesis and Triggered Drug Release. Journal of Materials Chemistry 2009;19(24) 4183-90.

174. Cheng R, Meng FH, Ma SB, Xu HF, Liu HY, Jing XB, Zhong Z. Reduction and Temperature Dual-Responsive Crosslinked Polymersomes for Targeted Intracellular Protein Delivery. Journal of Materials Chemistry 2011;21(47) 19013-20.

175. Dai J, Lin SD, Cheng D, Zou SY, Shuai XT. Interlayer-Crosslinked Micelle with Partially Hydrated Core Showing Reduction and pH Dual Sensitivity for Pin-Pointed Intracellular Drug Release. Angewandte Chemie 2011;50(40):9404-8.

176. Soppimath KS, Tan DCW, Yang YY. pH-Triggered Thermally Responsive Polymer Core-Shell Nanoparticles for Drug Delivery. Advanced Materials 2005;17(3) 318-23.

177. Zhang LY, Guo R, Yang M, Jiang XQ, Liu BR. Thermo and pH Dual-Responsive Nanoparticles for Anti-Cancer Drug Delivery. Advanced Materials 2007;19(19) 2988-92

178. Lo CL, Lin KM, Hsiue GH. Preparation and Characterization of Intelligent Core-Shell Nanoparticles based on Poly(D, L-lactide)-g-Poly(N-Isopropyl-Acryl-Amide-Co-Methacrylic Acid). Journal of Controlled Release 2005;104(3) 477-88.

179. Li W, Li JF, Gao J, Li BH, Xia Y, Meng YC, Yu Y, Chen H, Dai J, Wang H, Guo Y. The Fine-Tuning of Thermosensitive and Degradable Polymer Micelles for Enhancing Intracellular Uptake and Drug Release in tumors. Biomaterials 2011;32(15) 3832-44.

180. Xing ZM, Wang CL, Yan J, Zhang L, Li L, Zha LS. Dual stimuli responsive hollow nanogels with IPN structure for temperature controlling drug loading and pH triggering drug release. Soft Matter 2011;7(18) 7992-7.

181. Zhang JC, Wu LL, Meng FH, Wang ZJ, Deng C, Liu H, Zhong Z. pH and Reduction Dual-Bioresponsive Polymersomes for Efficient Intracellular Protein Delivery. Langmuir 2011;28(4) 2056-65.

182. Bahadur KCR, Thapa B, Xu P. pH and Redox Dual Responsive Nanoparticle for Nuclear Targeted Drug Delivery. Molecular Pharmaceutics 2012;9(9) 2719-29.

183. Chen J, Qiu XZ, Ouyang J, Kong JM, Zhong W, Xing MMQ. pH and Reduction Dual-Sensitive Copolymeric Micelles for Intracellular Doxorubicin Delivery. Biomacromolecules 2011;12(10) 3601-11.

184. Pan YJ, Chen YY, Wang DR, Wei C, Guo J, Lu DR, Chu CC, Wang CC. Redox/pH Dual Stimuli-Responsive Biodegradable Nanohydrogels with Varying Responses to Dithiothreitol and Glutathione for Controlled Drug Release. Biomaterials 2012; 33(27) 6570-9.

185. Shao Y, Huang W, Shi C, Atkinson ST, Luo J. Reversibly Crosslinked Nanocarriers for On-Demand Drug Delivery in Cancer Treatment. Therapeutic delivery 2012;3(12) 1409-27.

186. Barick KC, Singh S, Jadhav NV, Bahadur D, Pandey BN, Hassan PA. pH-Responsive Peptide Mimic Shell Cross-Linked Magnetic Nanocarriers for Combination Therapy. Advanced Functional Materials 2012;22(23) 4975-4.

187. Zhao ZH, Huang DT, Yin ZY, Chi XQ, Wang XM, Gao JH. Magnetite Nanoparticles as Smart Carriers to manipulate the Cytotoxicity of Anticancer Drugs: Agnetic Control and pH-Responsive Release. Journal of Materials Chemistry 2012;22(31) 15717-5.

188. Gan Q, Lu XY, Yuan Y, Qian JC, Zhou HJ, Lu X, Shi J, Liu C. A Magnetic, Reversible pH-Responsive Nanogated Ensemble based on Fe3O4 Nanoparticles-Capped Mesoporous Silica. Biomaterials 2011;32(7) 1932-2.

189. Gan Q, Lu XY, Dong WJ, Yuan Y, Qian JC, Li YS, Shi J, Liu C. Endosomal pH-Activatable Agnetic Nanoparticle-Capped Mesoporous Silica for Intracellular Controlled Release. Journal of Materials Chemistry 2012;22(31) 15960-8.

190. Xu HF, Meng FH, Zhong ZY. Reversibly Crosslinked Temperature-Responsive Nano-Sized Polymersomes: Synthesis and Triggered Drug Release. Journal of Materials Chemistry 2009;19(24) 4183-90.

191. Cheng R, Meng FH, Ma SB, Xu HF, Liu HY, Jing XB, Zhong Z. Reduction and Temperature Dual-Responsive Crosslinked Polymersomes for Targeted Intracellular Protein Delivery. Journal of Materials Chemistry 2011;21(47) 19013-20.

192. Du JZ, Du XJ, Mao CQ, Wang J. Tailor-Made Dual pH-Sensitive Polymer-Doxorubicin Nanoparticles for Efficient Anticancer Drug

Delivery. Journal of the American Chemical Society 2011;133(44) 17560-3.

193. Sankaranarayanan J, Mahmoud EA, Kim G, Morachis JM, Almutairi A. Multi-Response Strategies to Modulate Burst Degradation and Release from Nanoparticles. ACS Nano 2010;4(10) 5930-6.

194. Li YP, Xiao WW, Xiao K, Berti L, Luo J, Tseng HP, Fung G, Lam KS. Well-Defined, Reversible Boronate Crosslinked Nanocarriers for Targeted Drug Delivery in Response to Acidic pH Values AND Cis-Diols. Angewandte Chemie 2012;124(12) 2918-23.

195. Liu TY, Hu SH, Liu KH, Shaiu RS, Liu DM, Chen SY. Instantaneous Drug Delivery of Magnetic/Thermally Sensitive Nanospheres by a High-Frequency Magnetic Field. Langmuir 2008;24(23) 13306-11.

196. Chen C, Geng J, Pu F, Yang XJ, Ren JS, Qu XG. Polyvalent Nucleic Acid/Meso-Porous Silica Nanoparticle Conjugates: Dual Stimuli-Responsive Vehicles for Intracellular Drug Delivery. Angewandte Chemie 2011;123(4) 912-6.

197. Klaikherd A, Nagamani C, Thayumanavan S. Multi-Stimuli Sensitive Amphiphilic Block Copolymer Assemblies. Journal of American Chemical Society 2009;131(13) 4830-8.

198. Chang BS, Sha XY, Guo J, Jiao YF, Wang CC, Yang WL. Thermo and pH Dual Responsive, Polymer Shell Coated, Magnetic Mesoporous Silica Nanoparticles for Controlled Drug Release. Journal of Materials Chemistry 2011;21(25) 9239-47.

199. Bilalis P, Chatzipavlidis A, Tziveleka LA, Boukos N, Kordas G. Nanodesigned Magnetic Polymer Containers for Dual Stimuli Actuated Drug Controlled Release and Magnetic Hyperthermia Mediation. Journal of Materials Chemistry 2012;22(27) 13451-4.

200. Wang K, Guo DS, Wang X, Liu Y. Multistimuli Responsive Supramolecular Vesicles based on the Recognition of P-Sulfonatocalixarene and its Controllable Release of Doxorubicin. ACS Nano 2011;5(4) 2880-94.

201. Loh XJ, Jesus del B, Toh PPC, Lee TC, Jiao DZ, Rauwald U, Appel EA, Scherman OA. Triply Triggered Doxorubicin Release from Supramolecular Nanocontainers. Biomacromolecules 2012;13(1) 84-91.

202. Heller J, Self-regulated drug delivery systems. Medical Device and Diagnostic Industry 1985;7 32-37.

203. Zhao L, Xiao C, Ding J, He P, Tang Z, Pang X, Zhuang X, Chen X. Facile One-Pot Synthesis of Glucose-Sensitive Nanogel via Thiol-ene Click Chemistry for Self-Regulated Drug Delivery. Acta Biomaterialia 2013;9(5) 6535-43.

204. Heller J. Chemically Self-Regulated Drug Delivery Systems. Journal of Controlled Release 1988;8(2) 111-125.

205. Zhou G, Lu Y, Zhang H, Chen Y, Yu Y, Gao J, Sun D, Zhang G, Zou H, Zhong Y. A novel pulsed drug-delivery system: polyelectrolyte layer-by-layer coating of chitosan-alginate microgels. International Journal of Nanomedicine 2013;8 877-87.

206. Singh BN, Kim KH. Floating Drug Delivery Systems: An Approach to Oral Controlled Drug Delivery via Gastric Retention. Journal of Controlled Release 2000;63(3) 235-59.

207. Alhaique F, Marchetti M, Riccieri F, Santucci E. A Polymeric Film Responding in Diffusion Properties to Environmental PH Stimuli: A Model for a Self-Regulating Drug Delivery System. The Journal of pharmacy and pharmacology 1981;33(7) 413-6.

208. Alhaique F, Riccieri FM, Santucci E, Crescezi V, Gamini A. A Possible pH-Controlled Drug Delivery System based on a Derivative of the Polysaccharide Scleroglucan, Journal of Pharmacy and Pharmacology 1984;37(5) 310-313.

209. Bala K, Vasudevan P. pH-Sensitive Microcapsules for Drug Release, Journal of Pharmaceutical Sciences 1982;71(8) 960-962.

210. Okahata Y, Ozaki K, Seki T. pH-Sensitive Permeability Control of Polymer Grafted Nylon Capsule Membranes. Journal of the Chemical Society, Chemical Communications 1984;(8) 518-521.

211. Okahata Y, Seki T, pH-Sensitive Capsule Membranes. Reversible Permeability Control from the Dissociative Bilayer-Coated Capsule Membrane by an Ambient pH Change. Journal of American Chemical Society 1984;106(26) 8065–8070.

212. Siegel RA, Firestone BA. pH-Dependent Equilibrium Swelling Properties of Hydrophobic Polyelectrolyte Copolymer Gels. Macromolecules 1988;21(11) 3254-3259.

213. Heller J, Trescony PV. Controlled Drug Release by Polymer Dissolution. II: Enzyme-Mediated Delivery Device. Journal of Pharmaceutical Sciences 1979;68(7) 919-921.

214. Trehan A, Ali A. Recent Approaches in Insulin Delivery. Drug Development and Industrial Pharmacy 1998;24(7) 589-97.

215. Jeong SY, Kim SW, Eenink MJD, Feijen J. Self-Regulating Insulin Delivery Systems. I. Synthesis and Characterization of Glycosylated Insulin. Journal of Controlled Release 1984;1(1) 57-66.

216. Fischel-Ghodsian F, Brown L, Mathiowitz E, Brandenburg D, Langer R. Enzymatically Controlled Drug Delivery. Proceedings of the National Academy of Sciences of the United States of America 1988;85(7) 2403-2406.

217. Hassan CM, Doyle III FJ, Peppas NA.Dynamic Behavior of Glucose-Responsive Poly(methacrylic acid-g-ethylene glycol) Hydrogels. Macromolecules 1997;30 6166-6173.

218. Kost J, Horbett TA, Ratner BD, Singh M. Glucose Sensitive Membranes Containing Glucose Oxidase: Activity, Swelling and Permeability Studies. Journal of Biomedical Materials Research 1985;19(9) 1117-1133.

219. Albin G, Horbett TA,. Ratner BD. Glucose Sensitive Membranes for Controlled Delivery of Insulin: Insulin Transport Studies. Journal of Controlled Release 1985;2 153-164.

220. Jin X, Zhang X, Wu Z, Teng D, Zhang X, Wang Y, Wang Z, Li C. Amphiphilic Random Glycopolymer Based on Phenylboronic Acid: Synthesis, Characterization, and Potential as Glucose-Sensitive Matrix. Biomacromolecules 2009;10(6) 1337-45.

221. Ishihara K, Kobayashi M, Shonohara I. Insulin Permeation through Amphiphilic Polymer Membranes having 2-Hydroxyethyl Methacrylate Moiety. Polymer Journal 1984;16 647-651.

222. Ishihara K, Matsui K. Glucose Responsive Insulin Release from Polymer Capsule. Journal of Polymer Science Part C: Polymer Letters 1986;24(8) 413-417.

223. Ishihara K. Glucose-Responsive Polymers for Controlled Insulin Release. Proceedings International Symposium Controlled Release Bioact. Mater. 1988;15 168-169.

224. You JO, Auguste DT. Feedback-Regulated Paclitaxel Delivery based on Poly(N,N-Dimethylaminoethyl Methacrylate-Co-2-Hydroxyethyl Methacrylate) Nanoparticles. Biomaterials 2008;29(12) 1950-7.

225. Ludvig N, Medveczky G, Rizzolo R, Tang HM, Baptiste SL, Doyle WK, Devinsky O, Carlson C, French JA, Kral JG, Charchaflieh J, Kuzniecky RI. An Implantable Triple-Function Device for Local Drug Delivery, Cerebrospinal Fluid Removal and EEG Recording in the Cranial Subdural/Subarachnoid Space of Primates. Journal of Neuroscience Methods 2012;203(2) 275-83.

226. Hiratani H, Alvarez-Lorenzo C. The Nature of Backbone Monomers Determines the Performance of Imprinted Soft Contact Lenses as Timolol Drug Delivery Systems. Biomaterials 2004;25(6) 1105-13.

227. Duffy DJ, Das K, Hsu SL, Penelle J, Rotello VM, Stidham HD. Binding Efficiency and Transport Properties of Molecularly Imprinted Polymer Thin Films. Journal of the American Chemical Society 2002;124(28) 8290-6.

228. Kumar N. Handbook of Particulate Drug Delivery (Nanotechnology book Series). Valencia, CA, USA: American Scientific Publishers, 2008, vol. 1.

229. Yamamoto Y, Tsutsumi Y, Yoshioka Y, Nishibata T, Kobayashi K, Okamoto T, Mukai Y, Shimizu T, Nakagawa S, Nagata S, Mayumi T. Site-Specific PEGylation of a Lysine-Deficient TNF-Alpha with Full Bioactivity. Nature Biotechnology 2003;21(5) 546-552.

230. Farokhzad OC, Dimitrakov JD, Karp JM, Khademhosseini A, Freeman MR, Langer R. Drug Delivery Systems in Urology--Getting "Smarter". Urology 2006;68(3) 463-9.

231. Patel MP, Patel RR, Patel JK. Chitosan Mediated Targeted Drug Delivery System: A Review. Journal of Pharmaceutical Sciences 2010;13(4) 536-57.

232. Kishida A. A Site-Specific Polymeric Drug Carrier for Renal Disease Treatment. Trends in Pharmacological Sciences 2003;24(12) 611-3.

233. Liu L, Fishman ML, Kost J, Hicks KB. Pectin-Based Systems for Colon-Specific Drug Delivery via Oral Route. Biomaterials. 2003;24(19) 3333-43.

234. Shirota K, Kato Y, Suzuki K, Sugiyama Y. Characterization of Novel Kidney-Specific Delivery System using an Alkylglucoside Vector. Journal of Pharmacology and Experimental Therapeutics 2001;299(2) 459-67.

235. Heaton LL, Lopez E, Maini PK, Fricker MD, Jones NS. Advection, Diffusion, and Delivery over a Network. Physical Review. E, Statistical, Nonlinear, and Soft Matter Physics 2012;86(2 pt 1) 021905-10.

236. Watkins PK, Walker AB, Verschoor GL. Dynamical Monte Carlomodeling of Organic Solar Cells: The Dependence of Internal Quantum Efficiency on Morphology. Nano Letters 2005;5(9) 1814-1818.

237. Basu S, Gerchman Y, Collins CH, Arnold FH, Weiss R. A Synthetic Multicellular System for Programmed Pattern Formation. Nature 2005;434(7037) 1130-4.

238. Pierobon M, Akyildiz I. Capacity of a Diffusion-based Molecular Communication System with Channel Memory and Molecular Noise. IEEE Transactions on Information Theory 2013;59(2) 942-954.

239. Moore M, Enomoto A, Nakano T, Egashira R, Suda T, Kayasuga A, Kojima H, Sakakibara H, Oiwa K. A Design of a Molecular Communication System for Nanomachines using Molecular Motors. Pervasive Computing and Communications Workshops, 2006. PerCom Workshops 2006. Fourth Annual IEEE International Conference 2006; 6 p559.

240. Haas M, Mooletmar F, Meijer DK, de Zeeuw D. Specific Drug Delivery to the Kidney. Cardiovascular Drugs and Therapy 2002;16(6) 489-96.

241. Kitamura M, Fine LG. The Concept of Glomerular Self-Defense. Kidney International 1999;55(5) 1639-71.

242. Christensen EI, Bim H, Verroust P, Moestrup SK. Membrane Receptors For Endocytosis in the Renal Proximal Tubule. International Review Of Cytology 1998;180 237-84.

243. Leung S, Bendayan R. Role of P-Glycoprotein in the Renal Transport of Dideoxynucleoside Analog Drugs. Canadian Journal of Physiology and Pharmacology 1999;77(8) 625-30.

244. van Ginneken CA, Russel FG. Saturable Pharmacokineticsin the Renal Excretion of Drugs. Clinical Pharmacokinetics 1989;16(1) 38-54.

245. Brenner BM, Hostetter TH, Humes HD. Glomerular Permselectivity: Barrier Function based on Discrimination of Molecular Size and Charge. The American Journal of Physiology 1978;234(6) F455-F460.

246. Takakura Y, Fujita T, Hashida M, Sezaki H. Disposition Characteristics of Macromolecules in Tumor-Bearing Mice. Pharmaceutical Research 1990;7(4) 339-346.

247. Kamada H, Tsutsumi Y, Sato-Kamada K, Yamamoto Y, Yoshioka Y, Okamoto T, Nakagawa S, Nagata S, Mayumi T. Synthesis of a Poly(Vinylpyrrolidone-Co-Dimethyl Maleic Anhydride) Co-Polymer and its Application for Renal Drug Targeting. Nature Biotechnology 2003;21(4) 399-404.

248. Manil L, Davin JC, Duchenne C, Kubiak C, Foidart J, Couvreur P, Mahieu P. Uptake of Nanoparticles by Rat Glomerular Mesangial Cells In Vivo and In Vitro. Pharmaceutical Research 1994;11(8) 1160-5.

249. Choi CH, Zuckerman JE, Webster P, Davis ME. Targeting Kidney Mesangium by Nanoparticles of Defined Size. Proceedings of the National Academy of Sciences of the United States of America 2011;108(16) 6656-61.

250. Subbiah R, Ramalingam P, Ramasundaram S, Kim DY, Park K, Ramasamy MK, Choi KJ. N,N,N-Trimethyl Chitosan Nanoparticles for Controlled Intranasal Delivery of HBV Surface Antigen. Carbohydrate Polymers 2012;89(4) 1289-1297.

251. Brich Z, Ravel S, Kissel T, Fritsch J, Schoffmann A. Preparation and Characterization of a Water Soluble Dextran Immunoconjugate of Doxorubicin and the Monoclonal Antibody. Journal of Controlled Release 1992;19(1-3) 245-258.

252. Igarashi R, Mizushima Y, Takenaga M, Matsumoto K, Morizawa Y, Yasuda A. A Stable PGE1 Prodrug for Targeting Therapy. Journal of Controlled Release 1992;20(1) 37-46.

253. Kopecek J. Targetable Polymeric Anticancer Drugs. Temporal Control of Drug Activity. Annals of the New York Academy of Sciences 1991;618 335-344.

254. Nishikawa M, Ohtsubo Y, Ohno J, Fujita T, Koyama Y, Yamashita F, Hashida M, Sezaki H. Pharmacokinetics of Receptor-Mediated Hepatic Uptake of Glycosylated Albumin in Mice. International Journal of Pharmaceutics 1992;85(1-3) 75-85.

255. Yamaoka T, Tabata Y, Ikada Y. Fate of water-Soluble Polymers Administered via Different Routes. Journal of pharmaceutical sciences 1995;84(3) 349-354.

256. Yamaoka T, Tabata Y, Ikada Y. Distribution and Tissue Uptake of Poly (Ethylene Glycol) with Different Molecular Weights after Intravenous Administration to Mice. Journal of Pharmaceutical Sciences 1994;83(4) 601-606.

257. Wang G, Kuroda K, Enoki T, Grosberg A, Masamune S, Oya T, Takeoka Y, Tanaka T. Gel Catalysts That Switch On and Off. Proceedings of the National Academy of Sciences of the United States of America 2000;97(18) 9861-4.

258. Alvarez-Lorenzo C, Concheiro A. Molecularly Imprinted Polymers for Drug Delivery. Journal of chromatography. B, Analytical Technologies in the Biomedical and Life Sciences 2004;804(1) 231-45.

259. Guo Z, Pereira T, Choi O, Wang Y, Hahn HT. Surface Functionalized Alumina Nanoparticle Filled Polymeric Nanocomposites with Enhanced Mechanical Properties. Journal of Materials Chemistry 2006;16(1) 2800-2808.

260. Dreis S, Rothweiler F, Michaelis M, Cinatl J, Kreuter J, Langer K. Preparation, Characterization and Maintenance of Drug Efficacy of Doxorubicin-Loaded Human Serum Albumin (HSA) Nanoparticles. International Journal of Pharmaceutics 2007;341(1-2) 207-214.

261. Lee MK, Lim SJ, Kim CK. Preparation, Characterization and In Vitro Cytotoxicity of Paclitaxel-Loaded Sterically Stabilized Solid Lipid Nanoparticles. Biomaterials 2007;28(12) 2137-2146.

262. Lee ES, Na K, Bae YH. Polymeric Micelle for Tumor pH and Folate-Mediated Targeting. Journal of Controlled Release 2003;91(1-2) 103-113.

263. Wang J, Mongayt D, Torchilin VP. Polymeric Micelles for Delivery of Poorly Soluble Drugs: Preparation and Anticancer Activity In Vitro of Paclitaxel Incorporated into Mixed Micelles based on Poly(Ethylene Glycol)-Lipid Conjugate and Positively Charged Lipids. Journal of Drug Targeting 2005;13(1-2) 73-80.

264. Gao Z, Lukyanov A, Singhal A, Torchilin V. Diacyllipid-Polymer Micelles as Nanocarriers for Poorly Soluble Anticancer Drugs. Nano Letters 2002;2(9) 979-982.

265. Wong HL, Rauth AM, Bendayan R, Wu XY. In Vivo Evaluation of a New Polymer–Lipid Hybrid Nanoparticle (Pln) Formulation of

Doxorubicin in a Murine Solid Tumor Model. European Journal of Pharmaceutics and Biopharmaceutics 2007;65(3) 300-308.

266. Liu J, Zeng F, Allen C. Influence of Serum Protein on Polycarbonate-based Copolymer Micelles as a Delivery System for a Hydrophobic Anti-Cancer Agent. Journal of Controlled Release 2005;103(2) 481-497.

267. Latallo JF, Candido KA, Cao Z, Nigavekar SS, Majoros IJ, Thomas TP, Balogh LP, Khan MK, Baker J. Nanoparticles Targeting of Anticancer Drug Improves Therapeutic Response in Animal Model of Human Epithelial Cancer. Cancer Research 2005;65(12) 5317-5324.

268. Wosikowski K, Biedermann E, Rattel B, Breiter N, Jank P, Loser R, Jansen G, Peters GJ. In Vitro and In Vivo Antitumor Activity of Methotrexate Conjugated to Human Serum Albumin in Human Cancer Cells. Clinical Cancer Research 2003;9(5) 1917-1926.

269. Schnyder A, Krahenbuhl S, Drewe J, Huwyler J. Targeting of Daunomycin using Biotinylated Immunoliposomes: Pharmacokinetics, Tissue Distribution and In Vitro Pharmacological Effects. Journal of Drug Targeting 2005;13(5) 325-335.

270. Choi SW, Kim JH. Design of Surface Modified Poly (D, L-Lactide-Co-Glycolide) Nanoparticles for Targeted Drug Delivery to Bone. Journal of Controlled Release 2007;122(1) 24-30.

271. Oh KT, Oh YT, Oh NM, Kim K, Lee DH, Lee ES. A Smart Flower-Like Polymeric Micelle for PH-Triggered Anticancer Drug Release. International Journal Of Pharmaceutics 2009;375(1-2) 163-169.

272. Cheng J, Teply BA, Sherifi I, Sung J, Luther G, Gu FX, Levy-Nissenbaum E, Radovic-Moreno AF, Langer R, Farokhzad OC. Formulation of Functionalized PLGA-PEG Nanoparticles for In Vivo Targeted Drug Delivery. Biomaterials 2007;28(5) 869-876.

273. Vinogradov SV, Batrakova EV, Kabanov AV. Nanogels for Oligonucleotide Delivery to the Brain. Bioconjugate Chemistry 2004;15(1) 50-60.

274. Patil YB, Toti US, Khadir A, Ma L, Panyam J. Single-Step Surface Functionalization of Polymeric Nanoparticles for Targeted Drug Delivery. Biomaterials 2009;30(5) 859-866.

275. Messerschmidt SKE, Musyanovych A, Altvater M, Scheurich P, Pfizenmaier K, Landfester K, Kontermann RE. Targeted Lipid-Coated Nanoparticles: Delivery of Tumor Necrosis Factor-Functionalized

Particles to Tumor Cells. Journal of Controlled Release 2009;137(1) 69-77.

276. Farokhzad OC, Jon S, Khademhosseini A, Tran TT, LaVan A, Langer R. Nanoparticle-Aptamer Bioconjugates: A New Approach for Targeting Prostate Cancer Cells. Cancer Research 2004;64(21) 7668-7672.

277. Schiffelers RM, Ansari A, Xu J, Zhou Q, Tang Q, Storm G, Molema G, Lu PY, Scaria PV, Woodle MC. Cancer siRNA Therapy by Tumor Selective Delivery with Ligand-Targeted Sterically Stabilized Nanoparticle. Nucleic Acids Research 2004;32(19) e149-e150.

278. Olivier JC. Drug Transport to Brain with Targeted Nanoparticles. NeuroRx 2005;2(1) 108-119.

279. Cho CS, Cho KY, Park IK, Kim SH, Sasagawa T, Uchiyama M, Akaike T. Receptor-Mediated Delivery of All Trans-Retinoic Acid to Hepatocyte Using Poly(L-Lactic Acid) Nanoparticles Coated with Galactose-Carrying Polystyrene. Journal of Controlled Release 2001;77(1-2) 7-15.

280. Elamanchili P, Diwan M, Cao M, Samuel J. Characterization of Poly(D,L-Lactic-Co-Glycolic Acid) based Nanoparticulate System for Enhanced Delivery of Antigens to Dendritic Cells. Vaccine 2004;22(19) 2406-2412.

281. Zhang L, Gu FX, Chan JM, Wang AZ, Langer RS, Farokhzad OC. Nanoparticles in Medicine: Therapeutic Applications and Developments. Clinical Pharmacology and Therapeutics 2008;83(5) 761-769.

282. Roux S, Garcia B, Bridot JL, Salome M, Marquette C, Lemelle L, Gillet P, Blum L. Perriat P, Tillement O. Synthesis and Characterization of Dihydrolipoic Acid Capped Gold Nanoparticles and Functionalization by the Electroluminescent Luminal. Langmuir 2005;21(6) 2526-2536.

283. Zatats M, Katz E, Baron R, Willner I. Reconstitution of Apo Glucose Dehydrogenase on Pyrroloquinoline Quinine-Functionalized Au Nanoparticles Yields an Electrically Contacted Biocatalyst. Journal of the American Chemical Society 2005;127(35) 12400-12406.

284. Ulman A. Formation and Structure of Self-Assembled Monolayers. Chemical Reviews 1996;96(4) 1533-1554.

285. Fan H, Chen Z, Brinker CJ, Clawson J, Alam T. Synthesis of Organo-Silane Functionalized Nanocrystal Micelles and their Self-Assembly. Journal of the American Chemical Society 2005;127(40) 13746-13747.

286. Ramirez E, Jansat S, Philippot K, Lecante P, Gomez M, Masdeu-Bulto AM, Chaudret B. Influence of Organic Ligands on the Stabilization of Palladium Nanoparticles. Journal of Organometallic Chemistry 2004;689(24) 4601-4610.

287. Woehrle GH, Hutchison JE. Thiol-Functionalized Undecagold Clusters by Ligand Exchange: Synthesis, Mechanism, and Properties. Inorganic Chemistry 2005;44(18) 6149-6158.

288. Sandi G, Kizilel R, Carrado KA, Fernandez-saavedra R Castagnola N. Effect of the Silica Precursor on the Conductivity of Hectorite-Derived Polymer Nanocomposites. Electrochimica Acta 2005;50(19) 3891-3896.

289. Huang WY, Han CD. Dispersion Characteristics and Rheology of Organoclay Nanocomposites based on a Segmented Main-Chain Liquid-Crystalline Polymer Having Pendent Pyridyl Group. Macromolecules 2006;39(1) 257-267.

290. Mammeri F, Le Bourhis E, Rozes L, Sanchez CJ. Mechanical Properties of Hybrid Organic-Inorganic Materials. Journal of Materials Chemistry 2005;15(1) 3787-3811.

291. Sanchez C, Ribot F. Design of Hybrid Organic-Inorganic Materials Synthesized Via Sol-Gel Chemistry. New Journal of Chemistry 1994;18(1) 1007-1047.

292. Judeinstein P, Sanchez C. Hybrid Organic-Inorganic Materials: A Land of Multidisciplinarity. Journal of Materials Chemistry 1996;6(1) 511-525.

293. Zhang X, Simon LC. In Situ Polymerization of Hybrid Polyethylene-Alumina Nanocomposites. Macromolecular Materials and Engineering 2005;290(6) 573-583.

294. Gao SL, Mader E. Characterization of Interphase Nanoscale Property Variations in Glass Fibre Reinforced Polypropylene and Epoxy Resin Composites. Composites Part A: Applied Science and Manufacturing 2002;33(4) 559-576.

295. Shenhar R, Norsten TB, Rotello VM. Polymer-Mediated Nanoparticle Assembly: Structural Control and Applications. Advanced Materials 2005;17(6) 657-669.

296. Rothenfluh DA, Bermudez H, O'neil1 CP, Hubbell JA. Biofunctional Polymer Nanoparticles for Intra-Articular Targeting and Retention in Cartilage. Nature Materials 2008;7(3) 248-254.

297. Arap W, Kolonin MG, Trepel M, Lahdenranta J, Cardo-vila M, Giordano RJ, Mintz PJ, Ardelt PU, Yao VJ, Vidal CL, Flamm CL, Valtanen H, Weavind LM, Hicks ME, Pollock RE, Botz GH, Bucana CD, Koivunen E, Cahill D, Troncoso P, Baggerly KA, Pentz RD, Do KA, Logothetis CJ, Pasqualini R. Steps Toward Mapping the Human Vasculature by Phage Display. Nature medicine 2002;8(2) 121-127.

298. Kolonin MG, Sun J, Do KA, Vidal Cl, Ji Y, Baggerly KA, Pasqualini R, Arap W. Synchronous Selection of Homing Peptides for Multiple Tissues by In Vivo Phage Display. FASEB Journal 2006;20(7) 979-981.

299. Smith GP, Petrenko VA. Phage Display. Chemical Reviews 1997;97(2) 391-410.

300. Rehor A, Hubbell JA, Tirelli N. Oxidation-Sensitive Polymeric Nanoparticles. Langmuir 2005;21(1) 411-417.

301. Harris JM, Martin NE, Modi M. Pegylation: A Novel Process for Modifying Pharmacokinetics. Clinical Pharmacokinetics 2001;40(7) 539-551.

302. Orringer DA, Koo YE, Chen T, Kopelman R, Sagher O, Philbert MA. Small Solutions for Big Problems: The Application of Nanoparticles to Brain Tumor Diagnosis and Therapy. Clinical Pharmacology & Therapeutics 2009;85(5) 531-534.

303. Prescott JH, Lipka S, Baldwin S, Sheppard NF, Maloney JM, Coppeta J, Yomtov B, Staples MA Santini JT. Chronic, Programmed Polypeptide Delivery from an Implanted, Multireservoir Microchip Device. Nature Biotechnology 2006;24(4) 437-438.

304. Lutz JF. Solution Self-Assembly of Tailor-Made Macromolecular Building Blocks Prepared by Controlled Radical Polymerization Techniques. Polymer International 2006;55 979-993.

305. Jagur-Grodzinski J. Preparation of Functionalized Polymers using Living and Controlled Polymerizations. Reactive & Functional Polymers 2001;49(1) 1-54.

306. Matyjaszewski K. Macromolecular Engineering: From Rational Design through Precise Macromolecular Synthesis and Processing to Targeted Macroscopic Material Properties. Progress in Polymer Science 2005;30(8-9) 858-875.

307. Taton D, Gnanou Y, Matmour R, Angot S, Hou SJ, Francis R, Lepoittevin B, Moinard D, Babin J. Controlled Polymerizations as Tools

for the Design of Star-like and Dendrimer-like Polymers. Polymer International 2006;55(10) 1138-1145.

308. Brauncker WA, Matyjaszewski K. Controlled/Living Radical Polymerization: Features, Developments, and Perspectives. Progress in Polymer Science 2007;32(1) 93-146.

309. Zhang XW, Lian XM, Liu L, Zhang J, Zhao HY. Synthesis of Comb Copolymers with Pendant Chromophore Groups based on RAFT Polymerization and Click Chemistry and Formation of Electron Donor-Acceptor Supramolecules. Macromolecules 2008;41(21) 7863-7869.

310. Gondi SR, Vogt AP, Sumerlin BS. Versatile Pathway to Functional Telechelics via RAFT Polymerization and Click Chemistry. Macromolecules 2007;40(3) 474-481.

311. Gao H, Matyjaszewski K. Synthesis of Molecular Brushes by "Grafting Onto" Method: Combination of ATRP and Click Reactions. Journal of the American Chemical Society 2007;129(20) 6633-6639.

312. Liu H, Li CH, Liu HW, Liu SY. pH-Responsive Supramolecular Self-Assembly of Well-Defined Zwitterionic ABC Miktoarm Star Terpolymers. Langmuir 2009;25(8) 4724-4734.

313. Arora A, Hakim I, Baxter J, Rathnasingham R, Srinivasan R, Fletcher DA, Mitragotri S. Needle-Free Delivery of Macromolecules Across the Skin by Nano Liter Volume Pulsed Microjets. Proceedings of the National Academy of Sciences of the United States of America 2007;104(11) 4255-4260.

314. Bashir SL, Chew AL, Anigbogu A, Dreher F, Maibach HI. Physical and Physiological Effects of Stratum Corneum Tape Stripping. Skin Research Technology 2001;7(1) 40-48.

315. Doukas AG, Kollias N. Transdermal Drug Delivery with a Pressure Wave. Advanced Drug Delivery Reviews 2004;56(5) 559-579.

316. Habash RW, Bansal R, Krewski D, Alhafid HT. Thermal Therapy, Part 1: An Introduction to Thermal Therapy. Critical Reviews in Biomedical Engineering 2006;34(6) 459-489.

317. Karande P, Jain A, Mitragotri S. Discovery of Transdermal Penetration Enhancers by High-Throughput Screening. Nature Biotechnology 2004;22(2) 192-197.

318. Lee SH, Lee HH, Choi SS. Nanoparticle Popsicle: Transdermal Delivery of Nanoparticles using Polymeric Microneedle array. Korean Journal of Chemical Engineering 2011;28(9) 1913-1917.

319. Demir YK, Akan Z, Kerimoglu O. Characterization of Polymeric Microneedle Arrays for Transdermal Drug Delivery. 2013;8(10) e77289.

320. Tuan-Mahmood TM, McCrudden MT, Torrisi BM, McAlister E, Garland MJ, Singh TR, Donnelly RF. Microneedles for Intradermal and Transdermal Drug Delivery. European Journal of Pharmaceutical Sciences 2013;50(5) 623-37.

321. Prausnitz MR. Microneedles for Transdermal Drug Delivery. Advanced Drug Delivery Reviews 2004;56(5) 581-587.

322. Mitragotri S. Current Status and Future Prospects of Needle-Free Liquid Jet Injectors. Nature Reviews. Drug Discovery 2006;5(7) 543-548.

323. Paliwal S, Menon GK, Mitragotri S. Low-Frequency Sonophoresis: Ultrastructural basis for Stratum Corneum Permeability Assessed using Quantum Dots. Journal of Investigative Dermatology 2006;126(5) 1095-1101.

324. Guy RH, Kalia YN, Delgado-Charro MB, Merino V, Lopez A, Marro D. Iontophoresis: Electrorepulsion and Electroosmosis. Journal of Controlled Release 2000;64(1-3) 129-132.

325. Al-Qadi S, Grenha A, Remunan-Lopez C. Microspheres Loaded with Polysaccharide Nanoparticles for Pulmonary Delivery: Preparation, Structure and Surface Analysis. Carbohydrate Polymers 2011;86(1) 25-34.

326. Taranejoo S, Janmaleki M, Rafienia M, Kamali M, Mansouri M. Chitosan Microparticles Loaded with Exotoxin A Subunit Antigen for Intranasal Vaccination Against Pseudomonas Aeruginosa: An In Vitro Study. Carbohydrate Polymers 2011;83(4) 1854-1861.

Chapter 2

Synthesis and Characterization of Various Protein α-Lactalbumin Nanotubes Structures by Chemical Hydrolysis Method

Pegah Esmaeilzadeh[1], Zahra Fakhroueian[2*], Pouriya Esmaeilzadeh[3], Narges Mohammadi[4]

[1]Biomedical Material Department, Institute of Pharmacy, Martin Luther University, Halle-Wittenberg, Germany

[2]College of Chemical Engineering, University of Tehran, Tehran, Iran

[3]Process Simulation and Control Lab, Chemical Engineering Collage, Iran University of Science

and Technology (IUST), Tehran, Iran

[4]Department of Chemistry, College of Science, Alzahra University, Tehran. Iran

ABSTRACT

New water-based nanofluids including unparalleled milk protein α-lactalbumin hollow nano-bio-tubes using low cost, available and advanced partial chemical hydrolysis strategy in bottom-up nano-assembly have been employed in this work. The aqueous sol-gel chemistry in nanotechnology which we selected for this goal offers new fabrication as interesting smart protein nanotubes. The kinds of nanometer sized tubular structures such as waved, helically coiled, bent, bamboo-shaped, bead-like and branched single-walled protein nanotubes (SWPNTs) with a range of 3 - 8 nm in outer diameters were produced by this method. Complete characterization for natural produced nanotubes including SEM, TEM images, G bond and D bond in Raman spectroscopy, XRD patterns, DLS (Dynamic Light Scattering) and FTIR analysis were evaluated which they are most significant experiments in synthesized protein nanotubes soluble in clear water nanofluids and stabilization of transparent nanofluids was proved within more than one year after preparation. Various necessary ligand ion salts such as Mn^{2+}, Zn^{2+} and Ca^{2+} or mixtures as bridge makers and producing biological selfassembly hollow SWPNTs were performed and we focused on new chemical technology under specific acidic hydrolysis method not conventional enzymatic proteolysis and applying surfactants, pH reagent, Tris-HCl buffer, polar solvent which could be produced by β-sheet stacked hydrolysed protein α-lactalbumin mechanism under appropriate conditions to achieving high efficiency new protein nanotubes skeleton. They can be promising materials applied in food science, diet nutrition, nanomedicine, nano-biotechnology and surgery.

KEYWORDS

Nanobiotechnology; Milk Protein Nanotubes; Helical Self-Assembly; Water-Based Nanofluids; Chemical Hydrolysis; Nanomedicine

1. INTRODUCTION

Besides widespread advancements in synthesis of nanotubes from non-biological materials (mainly carbon nanotubes) attempts to mimic the ingeniously natural bottom-up nanofabrication with the aim of achieving the design and synthesis of man-made natural nanotubes, open the door to a new technology of important applications in the scope of nanobiotechnology. Natural bimolecular assemblies exhibit the selectivity and specificity which can adopt a wide variety of structures that perform an array of functions, and finally they are based on various chemistries (nucleic acids, peptides and proteins, lipids, and carbohydrates) that promote self-assembly in water at neutral pH and ambient temperatures [1]. On the road to these emerging opportunities, proteins represent fertile territory for nanobiotechnology, because they have properties ideal for nano-engineering purposes [2].The building blocks self-assemble and in the presence of ion Ca^{2+} salt bridges connect the individual building blocks, resulting in a hollow structure that previously, the kinetics of nanotube formation has been studied as a function of Ca^{2+} concentration. The most important functional properties of whey proteins are solubility, viscosity, gel formation, emulsification, foaming, and like many other albumins, the capacity to form nanoparticles (Mehravar et al., 2009) [3] and (H. Hernandez-Sanchez et al., 2012) could obtain nanoparticles of bovine α-lactalbumin α-LA products in the 100 to 200 nm size range so that they could be used as carriers of bioactive compounds [4]. Proteins also are the fundamental building blocks of all biological matter that exhibit such finesse in architectural design [5]. In just one reviewin 2006, J. F. GravelandBikker, et al., succeeded to create and synthesis novel alpha-lactalbumin (a α/β globular milk protein) nanotubesvia enzymatic proteolysis in presence of Bacillus licheniformis protease (BLP) for the first time which selfassembled into nanometer-sized tubular structures at appropriate conditions. Linear assemblies such as rod, fiber and tube-like structures are particular interest, because they have unique properties with respect to applications [6] and these findings represent important characteristic for the development of a nanotechnological application. Proteins and peptides can be synthesized by chemical or biological means. In this regard S. Banta et al. in 2007 researched about engineering of protein and peptide building blocks for nanotechnology usage [7] and the interactions that lead to the formation of these structures include chiral dipole-dipole interactions, π-π stacking, hydrogen bonds, nonspecific Van der Waals interactions, hydrophobic forces, electrostatic interactions, and repulsive steric forces [8]. Biological molecules such as peptides and proteins undergo self-assembly processes in vivo and in vitro, and these monomers are assembled into various nanometer-scale structures at room temperatures and atmospheric pressure [9]. The self-assembled protein and peptide nanostructures can be further organized to form various nanowires, nanotubes, and nanoparticles via their molecular-recognition functions. It seems that, the most special feature of α-lactalbumin protein nanotubes is due to its cavities and hollow cores. Because of their cavities, they could well serve as vehicles for drug or other encapsulated molecules such as vitamins and enzymesand also for doping of diverse ligand ions like Fe, Mn, Ca and Zn melts encapsulated in diet nutrition. During the headway of this research authors perceived that the unique stupendous linear self-assembly of α-lactalbumin building blocks in specific nanofluids using chemical hydrolysis and sol-gel matrix derived nanomaterials along with the matchless power of self-assembly can create thermodynamically stable nano-bio-tubes and rods products [10- 12]. This strategy also represents low cost example of bottom-up assembly and chemistry is entering the field of nanoscale science which is originally physics-oriented. This interesting well-orchestrated mechanism offers obvious advantages as the smart protein nanostructures as well as new authors' researches in this activated field. In this work we donot need touse any special catalyst, and we did not applyany enzymatic proteolysis [13,14] system and these green obtained protein nanofluids were fabricated by chemical sol-gel method as nanofluids in water solutions using bovine α-lactalbumin protein as precursor which hydrolyses under various conditions which it is a platform on nanomedicine. They showed very clear and stable protein cylinder nanotubes after one year, confirmed by DLS measurement and TEM images.

2. EXPERIMENTAL DETAILS

2.1. Materials

Bovine holo α-Lactalbumin (L6010) Sigma, Tris hydroxyl methyl aminomethane, acetyl acetone (acac), urea, citric acid, HCl 37%, Triethanolamine (TEA), calcium chloride ($2H_2O$), manganese chloride ($4H_2O$), zinc acetate, pure isopropyl alcohol. All chemicals were supplied from Merck, and domestic deionized water was obtained in this work.

2.2. Methods

Sample 1) The mixture of acacas nonionic surfactant, 10 ml isopropyl alcohol, 10 ml deionized water, 0.5 - 2 mg $MnCl_2 \cdot 4H_2O$ as connector and 0.5 - 2 mg protein α-lactalbumin was adjusted to acidic pH 2 - 3 using citric acid 1 M, HCl 2 M and urea 1 M solutions. The reaction solution was held in shaking water bath (Grant, OLS 200 from U.K.) at 42°C with 60 - 80 rpm during 1 - 2 weeks and then was filtered using 0.1 μl filter paper diameter (Millipore company). Finally the filtrate solution was held at 42°C during one week again and was kept at 4°C in refrigerator about 8 h (the consideration of acidic pH, Mn^{2+} ion metal at 42°C).

Sample 2) The mixtures of nonionic acac, isopropyl alcohol in deionized water were maintained at alkaline pH with aid of TEA (triethanolamine) as cationic surfactant and Tris-HCl buffer (pH 8 - 9, 75 mM) solutions. Then 100 μl from this solution with 100 μl of 0.5 - 2 mg protein α-lactalbumin dissolved in deionized water were mixed and eventually 0.001 - 0.005 mg of zinc acetate salt[Zn(CH3COO) $_2$]as metal ion salt was added. It was kept during 2 - 3 days in incubator at 37˚C and was stored at 4˚C for 3 days (alkaline pH, Zn^{2+} ion at 37˚C).

Sample 3) First of all, we should prepare the mixture of 3-5ml nonionic cacac with 20 ml isopropyl alcohol in deionized water. This precursor would be able to regulate in final neutral pH by urea 1M solution (pH controllerreagent) and Tris-HCl buffer (pH 8 - 9, 0.075 M) solution (Tris hydroxyl methyl amino methane a cationic surfactant) with mild stir up. Then ion salts solution of $MnCl_2{\cdot}4H_2O$ or $CaCl_2{\cdot}H_2O$ as the self-assembly essential stimulant motor was added to them. In the end of work, 0.5 - 2 mg protein α-lactalbumin solution was mixed. This product was kept between one-two weeks in shaking water bath at 42˚C and 60 - 80 rpm (mixed Mn + Ca ions, alkaline pH, at 42˚C).

Sample 4) 5 - 7 ml urea (1 M) with 4 ml isopropyl alcohol and 4 ml deionized water were mixed well together. After adding protein α-lactalbumin and $MnCl_2{\cdot}4H_2O$ the pH of solution adjust in final alkali pH by 0.075 M Tris hydroxyl methyl amino methane. The obtained reaction solution was incubated at 37˚C between one or two weeks and then transferred to 4˚C for almost one day (alkaline pH, Mn^{2+} ion, 37˚C, without acac).

Sample 5) Chemical hydrolysis solution was designed and prepared for use of 10 ml isopropyl alcohol, 0.075 M Tris-HCl buffer (pH 7.5), acac (acetyl acetone) and 4 - 6 ml one molar urea solution, the pH was adjusted (6.0 to 7.5). 200 μl of this stock solution was added to 200 μl from 0.5 - 2 mg protein α-lactalbumin solution. Then $MnCl_2{\cdot}4H_2O$ as a ligand forming ion metal was added, and the hydrolysis reaction was performed. The final solution was incubated for 3 - 5 days in incubator at 37˚C and stored at 4˚C in one day (neutral pH, Mn^{2+} ion, 37˚C).

2.3. Sample Preparation for Transmission Electron Microscopy (TEM)

Final solusions were taken into Ultrasound for 5 - 10 min, then 15 μL of the samples were placed on carbon coated grids with help of micropipet and be allowed to airdrying. Examination was done using a Philips TEM CM 120.

3. RESULTS AND DISCUSSION

3.1. SEM, TEM, XRD, Raman, DLS and FTIR Characterizations for Protein Nanotubes

Natural milk protein nanotubes characterized by SEM images and nanotubes morphology was observed and TEM pictures showed actually diameter of these nanotubes. XRD pattern indicated special structures according standard card and Raman spectroscopy investigated D and G bonds for exclusive nanotubes. Repeatability and stability of synthetic nanotubes structures in transparent water based solution was confirmed byqualitative dynamic light scattering (DLS) analysis and distribution size curves after long times from making. Doping of divalent ions-bridging between specific amino acids and carboxyl groups and finally the special vibrational bands for milk protein nanotubes are so considered by FTIR spectroscopy.

3.2. Characterization of Single Walled Protein Nanotubes Sample 1

In combined **Figure 1** SEM (Philips, XL30) and TEM (Philips, CM 120) images and also FTIR spectroscopy (Thermo Nicolet Nexus 870, USA) for sample 1 exhibitedvery complete natural single walled protein nanotubes (5 - 7.5 nm). The critical calculation predicted that various forms of helically coiled structures are possible and those structures are energetically and thermo dynamically stable. Helical and spring structures of carbon nanotubes may have been observed as early as the 1950s as filamentary carbon structures, or fibers [15].

TEM images show very stable protein nanotubes soluble after 3 months. FTIR spectrum shows very sharp transmission band centered at 1724.49 cm^{-1} should be attributed to the C=O ketone or carboxylic acid stretch and 1624.88 cm^{-1} frequency for C=O amide or contaminating water in carbon skeleton in PNTs structure respectively. The signal at 2921.52 cm^{-1} is attributed to $-CH_3$stretching vibration band. The peak at 1411.22 cm^{-1} can be assigned to stretching vibration of C=C in aromatic rings, the medium peaks at 1280.30 cm^{-1} and short peaks in range of 615.79 - 544.84 cm^{-1} are belong to special PNTs skeleton. A very weak signal at 825.92 cm^{-1} can be assigned to stretching vibration of C-O-C groups andthe weak peak at 1151.41 cm^{-1} can be interpreted to the characteristic of primary alcohol [16]. In addition, the very strong and sharp transmission band centered at 431.73 cm^{-1} may be for ion salt Mn^{2+} as bridging complex ligand.

3.3. Characterization of SWPNTs Sample 2 by SEM and TEM

SWPNTs (4 - 6 nm) for sample 2 was illustrated in **Figure 2** through SEM images of regularly bent during growth which are able to display the breaking of very thin film layers in boundary of nanofluids and exposured nano products from down over up. TEM images of branched products confirmed nearly 4 - 6 nm diameters for single-walled protein nanotubes.

In this important evidence, we can suggest the sharp G-bond and broad D-bond in limited ranges in **Figure 3**, because they can be different of pure carbon nanotubes (CNTs) and pure protein too.

3.4. Characterization of Beaded SWPNTs Sample 3 in Mixtures of Mn + Ca and Urea

We could prove that this product was kept stable after 3 months, by repetition of its Raman spectroscopy. (a) SEM, (b) TEM, (c) XRD(Seifert, Germany, 3003 PTS) and (d) FTIR were shown in**Figure 4**.

SEM and TEM images confirm the existence of very nice PNTs products soluble in nanofluid with diameter of (6.5 - 7.5) nm for SWPNTs. The XRD pattern of PNTs showed max intensity d_{100} plane in 21.5° (2θ) and another one in 70.73° (2θ) which is similar to CNTs. These results have made the Raman spectrumsuch a good tool to characterize PNTs that could prove these proofs. G-bond indicated at limited ranges between 1590.18 cm^{-1} (strong), –1427.01 cm^{-1} (short), and D-bond was seen as a weak bond at 1368.97 cm^{-1} (**Figure 5**(e)). All of these peaks have shifts related to CNTs. We could observe same peaks after 3 months in Raman spectroscopy (**Figure 5**(f)) which it showed very stability during this period. This morphology for sample 3 would produce under same conditions with Ca^{2+} cation ligand too. The characteristic vibrational modes of PNTs are: strong triplet peaks at 3464.67, 3361.50 and 3213.04 cm^{-1} for –OH and –NH stretching vibration modes, 1679.23 $cm^{-1}$for C=O stretching amide, 1625.44 cm^{-1} is attributed to contaminating water, the vibrational sharp signal centered at 1451.67 cm^{-1} is connected with the –OH bending deformation in carboxylic acids and phenolic groups and also it is belongs to –C=C– symmetric stretching vibration in CNTs. Two short and low broad peaks centered at 587.01 and 528.89 cm^{-1} can be accordance to inorganic bonds (Mn^{2+}-O) or (Ca^{2+}-O) ions metal in product. The extra peaks at around 878.04 - 786.82 cm^{-1} can be assigned to stretching vibration of C-O-C groups or specific of CNTsPNTs [17]. It was found that all of these PNTs vibrational modes have depicted fewer amounts of replacement frequencies and shifts in infrared spectra in comparison of CNTs.

3.5. Dynamic Light Scattering (DLS) Technique Analysis for Stable Nanofluid of Sample 3

The prepared protein nanofluids of product 3 can be maintained higher than 10 - 11 months (one year), which it was confirmed as only qualitative manner by DLSmeasurement. Furthermore,**Figure 6** shows three curves which they were measured by Nano-ZS Malvern system, (a) displays the size distribution by number in which 100% of these nanotubes have diameter of 5.48 nm, (b) the curve of size distribution by intensity that can help us determine the diameter size of 6nm for 78% intensity with 0.8 nm width, and finally (c) shows size distribution by volume in which total of nanotubes have shown the diameter sizes of 5.75 nm with 0.97 nm width into 100% volume. On the other hand, the size distribution of these nanotubes has remained under 10 nm after one year in 4°C, and these documents are very remarkable in performance of them.

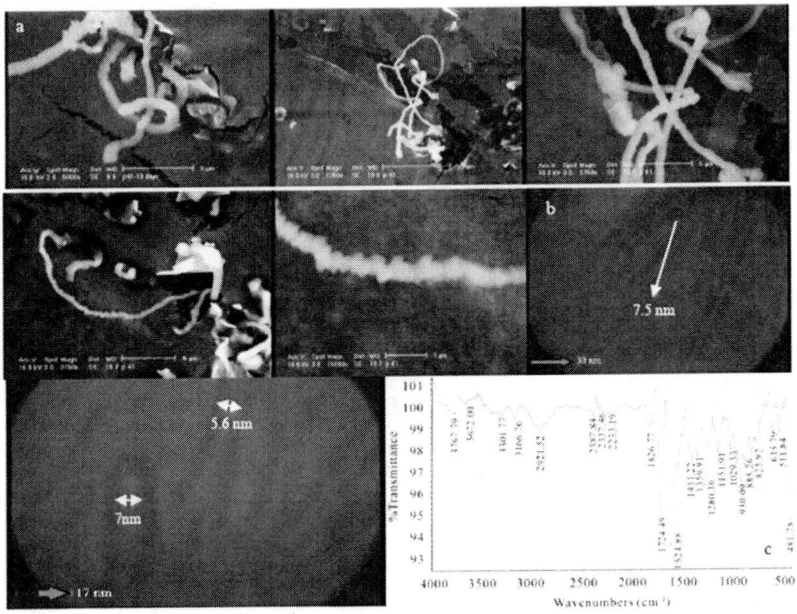

Figure 1. SEM (a) images of new helically coiled, (b) images of branched protein nanotubes, and FTIR spectroscopy (c) of SWNTs (5 - 7.5 nm) of PNTs.

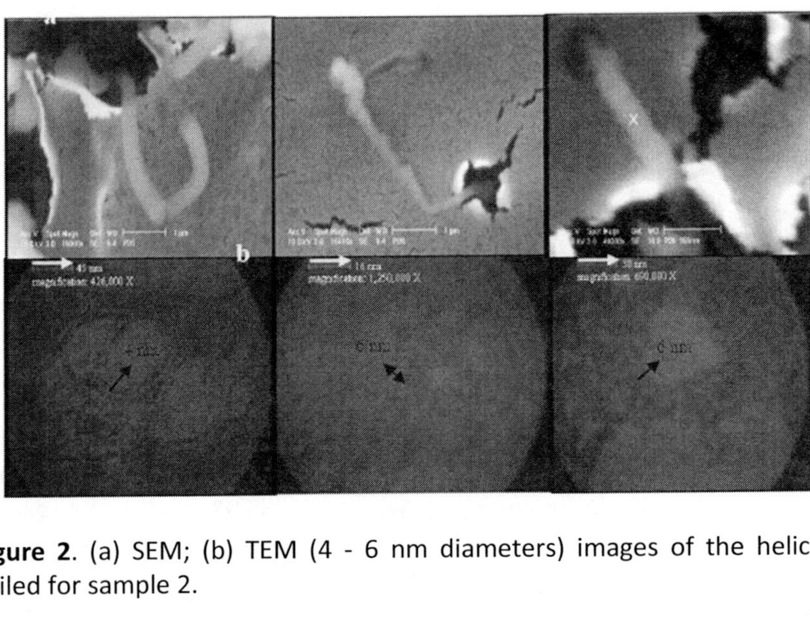

Figure 2. (a) SEM; (b) TEM (4 - 6 nm diameters) images of the helically coiled for sample 2.

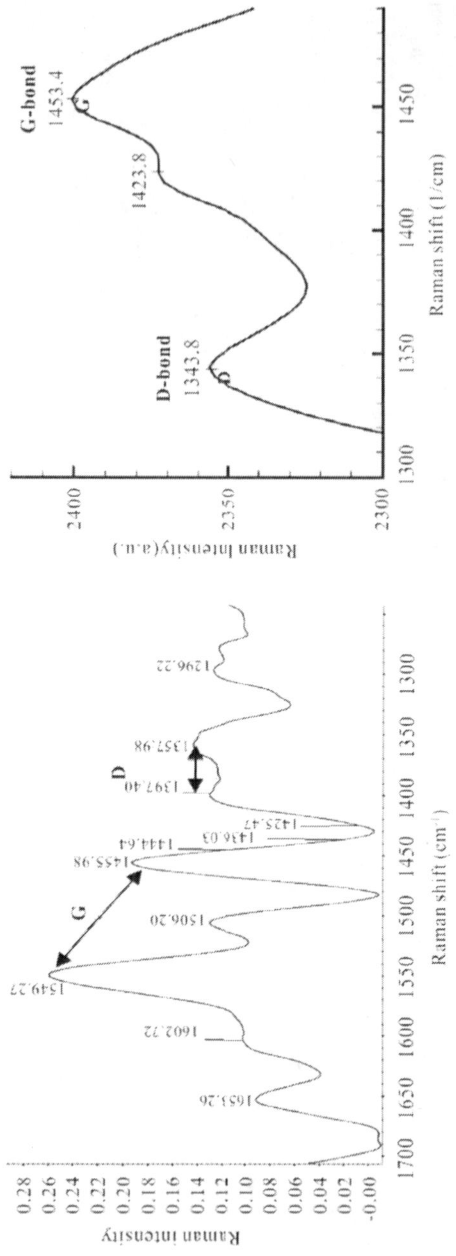

Figure 3. (a) Raman spectrum for PNTs of sample 2. G-bond spectrum is observed as doublet at ranges of (1549.27 - 1455.98 cm^{-1}) and broad D-bond is showed at (1397.40 - 1357.98 cm^{-1}); (b) G-bond (1453.4 cm^{-1}) and D bond (1343.8 cm^{-1}) in Raman spectroscopy for pure protein.

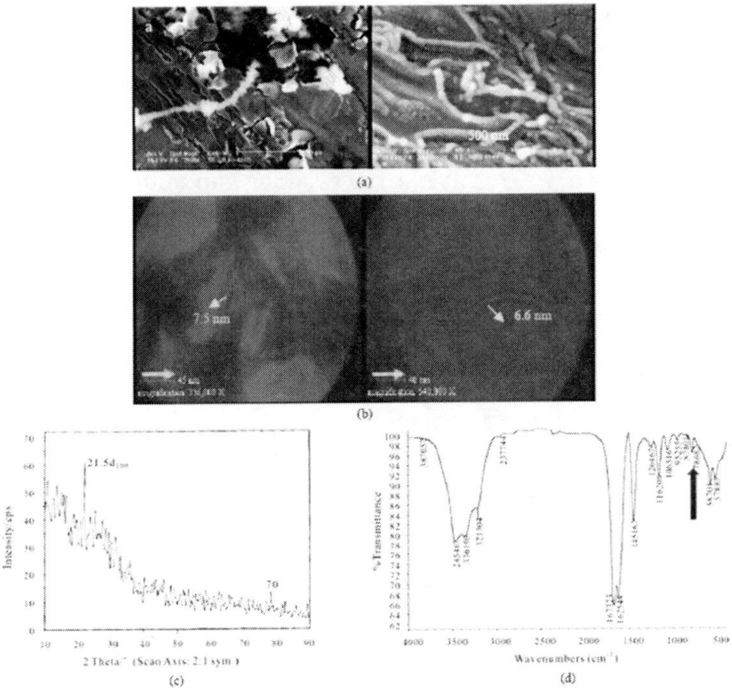

Figure 4. SEM and TEM pictures, XRD pattern and FTIR spectrum of bead-like single-walled protein nanotubes of sample 3 in appearance of two ion metal connectors.

3.6. Representation of Some Pictures of Protein Nanotubes of Product 3

Figure 7 shows schematic pictures of some PNTs nanofluid of product 3. (a) the evidence of beginning of synthesis and (b) physical transparency phenomenon after more than 11 months (one year) maintaining in 4°C.

3.7. Characterization of SWPNTs Sample 4 Containing Various Concentrations of Urea Reagent

The other final characterizations of natural protein nanotubes (NPNTs) in this research are about the production of single walled NPNTs for sample 4 which was shown in **Figure 8**. Bead-like nanotubes were fabricated by chemical hydrolysis including urea concentrations and alkaline pH value. The diameter of nanotubes was determined by TEM characterization in **Figure 8**(b).

SEM was measured at 37°C as narrow cylinder nanotubes of protein. TEM was shown this claim in diameters. XRD pattern determined all diffraction peaks which could be indexed in the protein nanotubes (PNTs) and also it compared with XRD of pure protein. The max sharp intensity for PNTs and broad pure protein curve were obtained in 22.17 (2θ) and 25 or 29.70 (2θ) respectively [18]. The strong and sharp peaks indicated that product has had good crystallinity. G-Bond (1515.03 cm^{-1}) and D-bond (1352.58 cm^{-1}) were also confirmed by Raman spectroscopy for these rolling amino acids nanostructures at 37°C. It seems that these new characterization are especial for PNTs. FTIR spectroscopy (d) is as well as sample 3 contains very sharp absorption at 564.22 cm^{-1} for in organic metal ion in PNTs structure.

3.8. Characterization of SWPNTs Sample 5 at pH 6 -7.5 after 3-5 days Annealing Time

Figure 9 shows new typical SEM images for protein nanotubes which they could grow as homogeneously perpendicular to the substrate like sweep-shape morphology (a). It is very interesting evidence, because the yield of synthetic products was reported very high and we can see them by TEM pictures (b). **Figure 10** indicates FTIR spectrum (c) which is similar to the sample 4. It consists of two-branched and strong peaks belonged to –OH and –NH stretching bands at 3456.12 - 3360.87 cm^{-1}, another sharp absorption amide band (C=O) was seen at 1671.67 - 1628.49 cm^{-1}which they attribute to protein structure, and finally the special short peaks at 587.65 - 522.48 and 469.40 cm^{-1} that they can be assigned to inorganic ligand metal Mn^{2+} and perfect PNTs nanostructures. The XRD pattern (d) shows max intensity phase d_{100} in 22.5° (2θ) with relatively sharp and strong peak, the other peaks at 37, 46, and 71° contain very low intensities, which imply and prove nanofine of nanotubes products. Raman spectrum (e) can be consider as doublet feature G-bond at 1565.01 - 1418.97 cm^{-1} ranges which was produced from the high degree of symmetry and order of carbon materials. They were generally used to identify well-ordered CNTs [19]. Generally the intensity of the E_2g modes of graphitic materials are sharp and strong when the sample is highly crystalline, while disordered graphites and carbons show a feature around 1353.13 cm^{-1} as singlet short

D-bond. On the other hand, the single peak at 1353.13 cm^{-1} might come from symmetry-lowering effect or defect effect in new products.

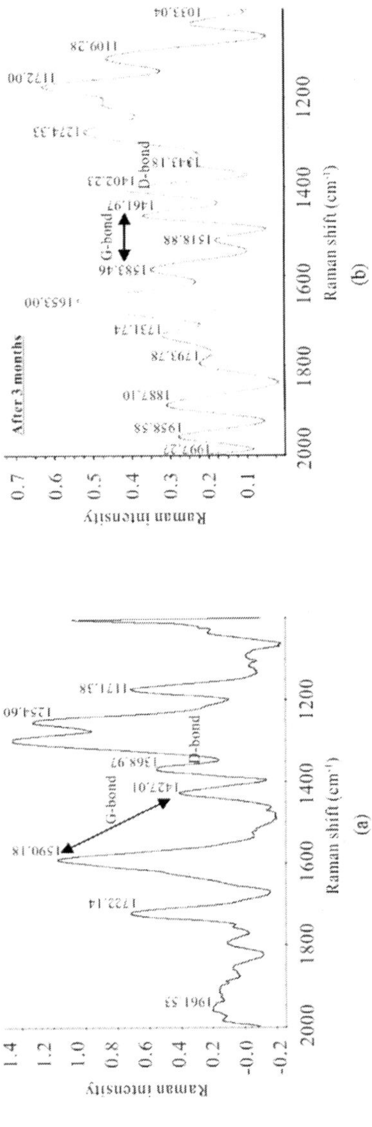

Figure 5. (a) and (b) Raman spectrum for PNTs of sample 3 (before and after 3 months). G-bond and D-bond spectra were observed at new frequencies, which they are different from of CNTs.

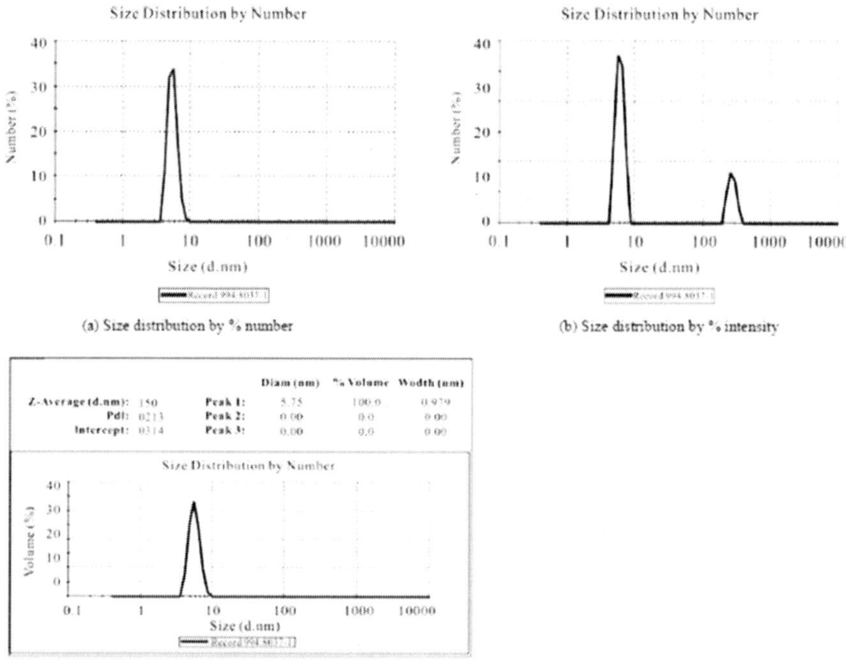

(a) Size distribution by % number

(b) Size distribution by % intensity

(c) Size distribution by % volume

Figure 6. The various DLS curves of qualitative size distribution of sample 3 by (a) % number; (b) % intensity; (c) % volume versus diameter size (nm).

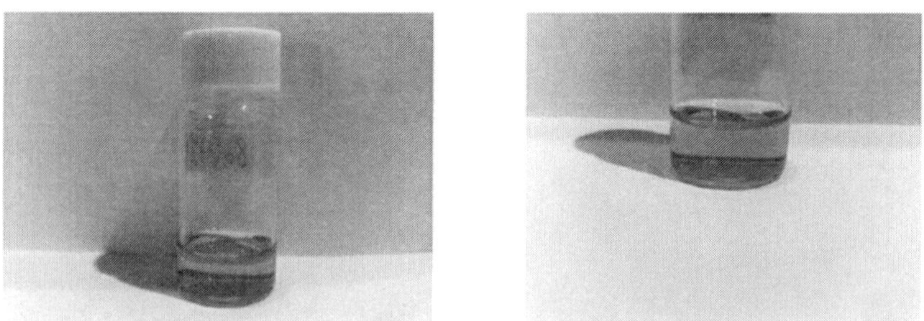

(a)

(b)

Figure 7. The exhibition of very clear and stability of α-lactalbum in nano-bio-tubes in two conditions, (a) in beginning of product synthesis and (b) after aging at 4˚C during one year.

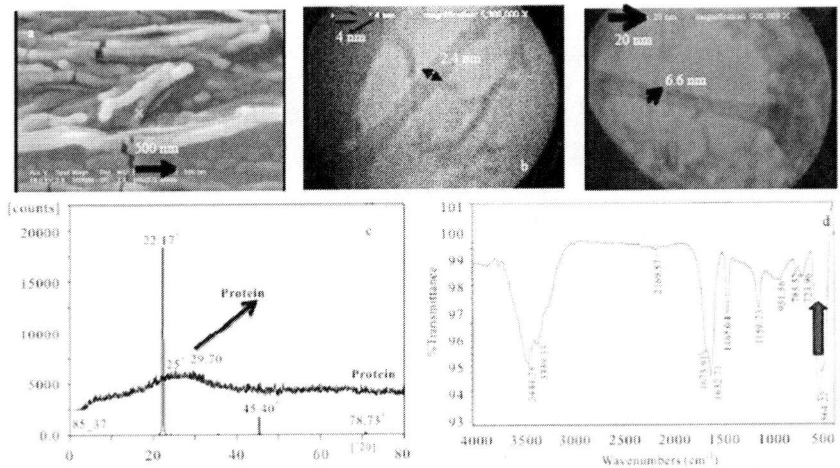

Figure 8. (a) SEM; (b) TEM images; (c) XRD pattern of nanotubes comparison with pure protein structure; (d) FTIR spectroscopy for synthetic SWPNTs sample 4.

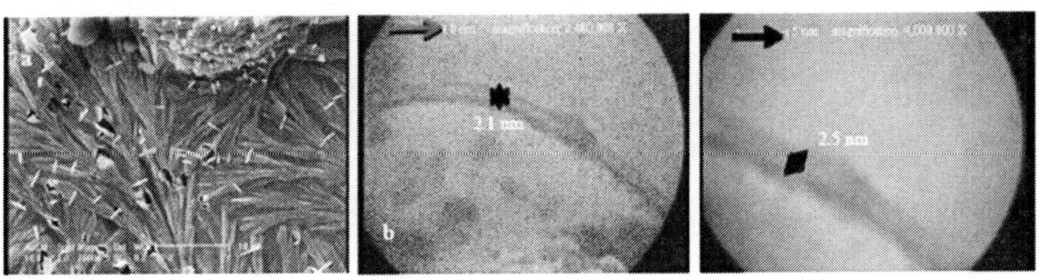

Figure 9. Typical new pictures of SWNPTs (3 nm) for PNTs, (a) SEM images; (b) TEM pictures of sample 5.

4. THE PROPOSED FORMATION MECHANISM OF PROTEIN NANOTUBES

Based on our difficult experiments in Lab., we found the optimum useful concentration and incubation temperatures and aging time for achievement of only various morphologies of protein nanotubes. Our suggested method should be able to hydrolysis of protein α-lactalbumin actually towards cleavage sites (13Asp-X and 7-Glu-X) of peptide bonds, then amino acids and carboxylates ions were degraded by chemical hydrolysis into nanoemulsion environment in atmospheric pressure. Various cations, such as divalent ions Mn^{2+}, Zn^{2+}, Ca^{2+} and Fe^{2+} or mixed of them can induce tubular self-assembly of partially hydrolysed α-lactalbumin and the molar cation ratio determines which one is faster than the other. Our aims are to do nucleation and growth mechanism towards helical regular structures. Apparently, they all possess a structural element that is required for helical self-assembly. Ions size ligands and coordination number are likely to play an additional role, the ion is required to act as a connector between two building blocks via a salt bridge. Then Ca^{2+} ion has a high affinity for the wellknown (primary) Ca^{2+} binding site and stabilizes the native conformation upon binding and a secondary binding site was located 0.8 nm away from the primary binding site [20]. The folding and subsequent assembly of polypeptide chains reaction between the molecules is believed to be the sum of β-sheet stacking involving Hbonds (β-sheet) and Ca^{2+} ion bridges between specific carboxylates and amino acids, fixing the molecules in the right orientation in the assembly [21]. Reliance on magic power of self-assembly, actually we just provide specific conditions via chemical hydrolysis and surfactants in nanoemulsions for produced protein units that they canthermo dynamically reach to the lowest free energy. Self-assembly phenomena binds head-to-tail of building blocks together, then they turn laterally selfassemble to make two-dimensional arrays that twist to form hollow cylindrical filaments, with a diameter of about 3 - 8 nm single walled nanotubes and finally stacking of ringshaped cyclic peptide subunits leads to protein nanotube formation [6]. Of course, it may have been made nanoparticles, nanorods, and many others shape, but we try to sophisticate and elaborate our proposal towards formation nanotubes [22].

5. CONCLUSION

Single-walled natural milk protein nanotubes including new morphologies were fabricated by a chemically direct rolling process that achieved in this work along with high efficiency, low cost and control ability over chemically sol-gel method, not by enzymatic proteolysis and without using individual special and traditional catalysts in synthesis and any need to high temperatures for the first time (in comparison with ordinary CNTs). These synthetic natural nanotubes with the property of multiple-responsiveness to temperature, pH, formulation of nanofluid, ligands binding, surfactants and aging time are envisioned to have immense applications, especially for preparing new medicine nanofluids or nutritional drugs containing clear water-based solutions consist of various Ca^{2+}, Zn^{2+} and Mn^{2+} ions in food science and pharmaceutics. So there are important points using Mn^{2+} and (Mn^{2+} + Ca^{2+}) at different pH that are able to produce helically coiled and bead-like nanotubes shapes in samples 1 and 3 (pH is major) and the effect of annealing times at two pH indicated very influence factors in formation of bottom-up structures in samples 4 and 5 (annealing time is important). The Zn^{2+} metal ion at alkaline pH could provide low yield branched production in sample 2.

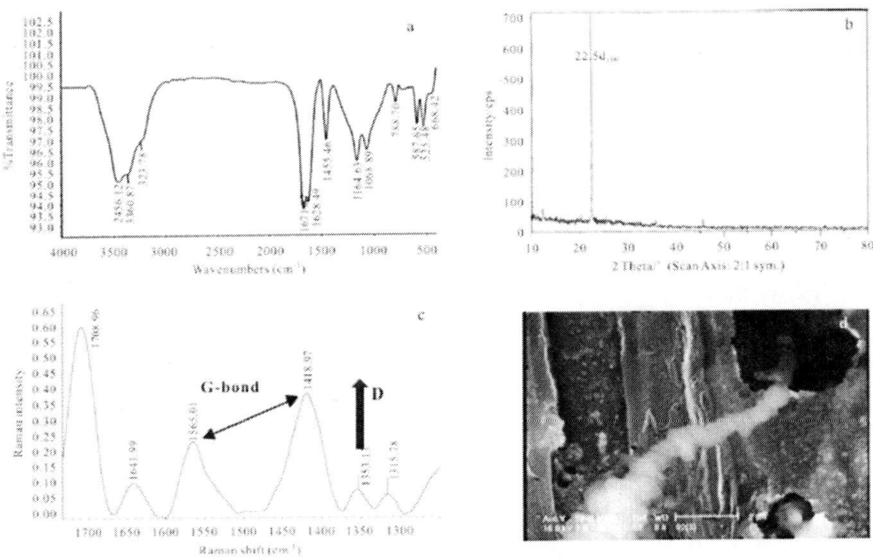

Figure 10. (a) FTIR spectroscopy; (b) XRD pattern; (c) Raman spectroscopy for sample 5; (d) shows SEM analysis of bamboo nanotubes form after 13 days keeping at 4°C.

ACKNOWLEDGEMENTS

The authors wish to express our sincere and deep gratitude to Mr. Rezai (SEM analysis), Dr. Sepehriseresht (TEM analysis), Mr. Naemi (XRD analysis), Mr. Ahmadi Taba (Raman and FTIR spectroscopy) and Mr. Bahari from RIPI (DLS analysis) for all vivacious times which we experienced by them.

REFERENCES

1. M. G. Ryadnov and D. N Woolfson, "Fiber Recruiting Peptides: Noncovalent Decoration of an Engineered Protein Scaffold," Journal of the American Chemical Society, Vol. 126, No. 24, 2004, pp. 7454-7455.

2. Y. Astie, H. Bayley and S. Howorka, "Protein Components for Nanodevices," Current Opinion in Chemical Biology, Vol. 9, No. 6, 2005, pp. 576-584.

3. R. Mehravar, M. Jahanshahi and N. Saghatoleslami, "Production of Biological Nanoparticles from α-Lactalbumin for Drug Delivery and Food Science Application," African Journal of Biotechnology, Vol. 8, No. 24, 2009, pp. 6822-6827.

4. I. J. Arroyo-Maya, J. O. Rodliles-Lopez, M. CornejoMazon, G. F. Gutierrez-Lopez, A. Hernandez-Arana, C. Toledo-Nunez, G. V. Barbosa-Canovas, J. O. FloresFlores and H. Hernandez-Sanchez, "Effect of Different Treatments on the Ability of α-Lactalbumin to Form Nanoparticles," Journal of Dairy Science, Vol. 95, No. 11, 2012, pp. 6204- 6214.

5. S. Keten and M. J. Buehler, "Large Deformation and Fracture Mechanics of a Beta-Helical Protein Nanotube: Atomistic and Continuum Modeling," Computer Methods in Applied Mechanics and Engineering, Vol. 197, No. 41, 2008, pp. 3203-3214.

6. J. F. Graveland-Bikker and C. G. de Kruifa, "Unique Milk Protein Based Nanotubes: Food and Nanotechnologymeet," Trends in Food Science & Technology, Vol. 17, No. 5, 2006, pp. 196-203.

7. S. Banta, Z. Megeed, M. Casali, K. Rege and M. L. Yarmush, "Engineering Protein and Peptide Building Blocks for Nanotechnology," Journal of Nanoscience and Nanotechnology, Vol. 7, No. 2, 2007, pp. 387-401.

8. J. D. Hartgerin and E. Beniash, S. Stupp, "Peptide-amphiphile nanofibers: A versatile scaffold for the preparation of self-assembling materials," Proceedings of the National Academy Sciences, Vol. 99, 2002, pp. 133-138.

9. X. Gao and H. Matsui, "Peptide-Based Nanotubes and Their Applications in Bionanotechnology," Advanced Materials, Vol. 17, No. 17, 2005, pp. 2037-2050.

10. P. Esmaeilzadeh, Z. Fakhroueian and A. A. Miran Beigi, "Synthesis of Biopolymeric α-Lactalbumin Protein Nanoparticles and Nanosphericals as Green Nanofluids Using in Drug Delivery and Food Technology," Journal of Nano Research, Vol. 16, 2011, pp. 89-96.

11. P. Esmaeilzadeh, Z. Fakhroueian, M. Jahanshahi, M. Chamani, H. R. Zamanizadeh and B. Rasekh, " A Synthetic Garden of State of the Art Natural Protein Nanoarchitectures Dispersed in Nanofluids," Journal of Biomedical Nanotechnology, Vol. 7, No. 3, 2011, pp. 1-8.

12. P. Esmaeilzadeh, Z. Fakhroueian and Po. Esmaeilzadeh, "Producing the Controlled Hybrid Lattice of Natural Protein Nanotubes by Chemical Hydrolysis for Nanomedicine," Journal of Nano Research, Vol. 21, 2013, pp. 7-13.

13. J. F. Graveland-Bikker, R, Ipsen, J. Otte and C. G. De Kruif, "Influence of Calcium on the Self-Assembly of Partially Hydrolyzed Alpha-Lactalbumin," Langmuir, Vol. 20, No. 16, 2004, pp. 6841-6846.

14. L. Chen, G. E. Remondetto and M. Subirade, "Food Protein-Based Materials as Nutraceutical Delivery Systems," Trend in Food Science and Technology, Vol. 17, No. 5, 2006, pp. 272-283.

15. M. Zhang and J. Li, "Carbon Nanotube in Different Shapes," Materials Today, Vol. 12, No. 6, 2009, pp. 12- 18.

16. L. Chen and H. Xie, "Surfactant-Free Nanofluids Containing Doubleand Single-Walled Carbon Nanotubes Functionalized by a Wet-Mechanochemicalreaction," Thermochemical Acta, Vol. 497, pp. 67-71.

17. G. Jiang, L. Wang, C. Chen, X. Dong, T. Chen and H. Yu, "Study on Attachment of Highly Branched Molecules onto Multiwalled Carbon Nanotubes," Material Letters, Vol. 59, No. 16, 2005, pp. 2085-2089.

18. X. Huang, C. Pan and X. Huang, "Preparation and Characterization of γ-MnO_2/CNTs Nanocomposite," Materials Letters, Vol. 61, No. 4-5, 2007, pp. 934-936.

19. D. Zhang, L. Shi, J. Fang, K. Dai and X. Li, "Preparation and Desalination Performance of Multiwall Carbon Nanotubes," Materials Chemistry and Physics, Vol. 97, No. 2-3, 2006, pp. 415-419.

20. M. Kronman, "Metal Ion Binding and the Molecular Conformational Properties of α-Lactalbumin," Critical Reviews in Biochemistry and Molecular Biology, Vol. 24, No. 6, 1989, pp. 565-667.

21. R. Ipsen and J. Otte, "Self-Assembly of Partially Hydrolysed α-Lactalbumin," Biotechnology Advances, Vol. 25, No. 6, 2007, pp. 602-605.

22. P. Esmaeilzadeh, Z. Fakhroueian, H. R. Zamanizadeh and B. Rasekh, "Single-Walled Natural Protein Nanotubes Soluble in Polar Solvent," International Journal of Green Nanotechnology, Vol. 3, No. 1, 2011, pp. 37-46.

Chapter 3

The impact of the confinement of reactants on the metal distribution in bimetallic nanoparticles synthesized in reverse micelles

Concha Tojo[1], Elena González[1] and Nuria Vila-Romeu[2]

[1]Physical Chemistry Department, Faculty of Chemistry, University of Vigo, 36310 Vigo, Spain

[2]Physical Chemistry Department, Faculty of Sciences, University of Vigo, 32004 Ourense, Spain

ABSTRACT

A kinetic study on the formation of bimetallic nanoparticles in microemulsions was carried out by computer simulation. A comprehensive analysis of the resulting nanostructures was performed regarding the influence of intermicellar exchange on reactivity. The objects of this study were metals having a difference in standard reduction potential of about 0.2–0.3 V. Relatively flexible microemulsions were employed and the concentration of the reactants was kept constant, while the reaction rate of each metal was monitored as a function of time using different reactant proportions. It was demonstrated that the reaction rates depend not only on the chemical reduction rate, but also on the intermicellar exchange rate. Furthermore, intermicellar exchange causes the accumulation of slower precursors inside the micelles, which favors chemical reduction. As a consequence, slower reduction rates strongly correlate with the number of reactants in this confined media. On the contrary, faster reduction rates are limited by the intermicellar exchange rate and not the number of reactants inside the micelles. As a result, different precursor proportions lead to different sequences of metal reduction, and thus the arrangement of the two metals in the nanostructure can be manipulated.

KEYWORDS

bimetallic nanoparticles; intermicellar exchange rate; microemulsion simulation

1. INTRODUCTION

The advancement in the field of nanotechnology relies on the improvement in nanoparticle preparation techniques. Thus, research has been targeted at the development of different synthetic technologies to achieve control over the composition, structure and shape of nanoparticles. Nowadays there are many methods available to synthesize nanoparticles. From the pioneering research of Boutonett et al. [1], synthesis using microemulsions became a frequently used technique [2-10]. This method offers many advantages with respect to other techniques, specifically, the possibility to prepare different types of materials using simple equipment while obtaining very small particles with a narrow size distribution whose composition is well-controlled. To synthesize nanoparticles by this method, one reactant is solved inside the droplets of a microemulsion, and another reactant is solved inside the droplets of a second microemulsion. After mixing, the microemulsions reactants can cross intermicellar channels and come into contact when they are located inside the same micelle due to micelle collisions and coalescence. The chemical reaction can then take place to form precipitates of nanometric size, which remain confined to the interior of reverse micelles. This approach has been used to prepare a variety of nanomaterials [6,11-15] that often display better catalytic activity than nano-catalysts prepared by other methods [10].

More complex structures, such as bimetallic nanoparticles, have also been prepared via microemulsions[11,14,16-19]. The combination of two different metallic atoms can result in the electronic coupling between the individual metals in the resulting particle, causing bimetallic nanoparticles to show different behavior from monometallic ones. The synthesis of bimetallic nanoparticles has drawn increasing attention in the field of nanotechnology, because their physical and chemical properties are composition-dependent [15,20-23]. Specifically, in the field of catalysis, bimetallic nanoparticles exhibit significantly increased catalytic behavior in comparison to monometallic nanoparticles [24]. Of interest is the controlled segregation and the extent of alloying of the two metals, since these factors profoundly affect catalytic activity and selectivity. Most of the catalytic reactions are structure sensitive, and bimetallic nanoparticles offer the opportunity to tune the catalytic properties by modifying the composition distribution [25]. Although it is of enormous importance to control the metal distribution in nanoparticles, predicting this is very difficult [26]. In addition, even in the case of bimetallic nanoparticles having similar compositions, different atomic distributions have been found depending on the preparation method. The accurate control of the bimetallic nanostructure remains a challenge. The design of catalysts for improved activity and selectivity must be based on a sufficiently good understanding of the reaction mechanism.

For a better insight into the metal distribution in a bimetallic nanoparticle synthesized in microemulsions, three aspects must be taken into consideration. First, it is believed that the nucleus develops into a particle by building up new layers, so the sequence of deposition of the metals determines the metal distribution. Second, this sequence is strongly dependent on the reduction rates of the two metal ions. Thus, if one metal reduces faster than the other, it will form the core. The metal which is reduced more slowly will be deposited onto this core, forming the outer layers, which results in a core–shell nanostructure. In contrast, a small difference between reduction rates of the two metals leads to an alloy. Third, this result was obtained in homogeneous reaction media. It was extended to synthesis inside micelles without taking into account the heterogeneity of reaction media. It is understood that reverse micelles are reaction vessels in which classical assumptions cannot always be used. For example, it was demonstrated that the concept as simple as the classical definition of pH cannot be applied in the interior of reverse micelles [27,28]. Previous simulation studies concluded that the structure of nanoparticles obtained in reverse micelles is determined by the difference in the reduction rates only if both reductions occur at the same rate or if reductions have very different rates [23,29]. That is, in these two extreme cases, compartmentalization of the reaction media cannot modify the metal arrangement. However, a vast majority of bimetallic systems belong to the large category between these extremes in which the metal segregation depends on the microemulsion dynamics.

With the exception of the existing studies relating nanoparticle properties to microemulsion composition [2,3,30-33], there is a gap regarding the impact of microemulsion dynamics on metal segregation in bimetallic nanoparticles. The synthesis of simple [34,35] and bimetallic [29,36] nanoparticles in microemulsions has been previously studied. The research at hand is focused on the rates of the chemical reaction inside micelles, which may provide fundamental knowledge on how the metal distribution in the nanostructure can be modified. The hypothesis is that the resulting nanostructure is due to the particular combination of three main factors: the intermicellar exchange rate (determined by microemulsion composition), the difference in reduction rate of both metals, and the amount of metal precursors inside the micelles. Together, this leads to a particular sequence of deposition of the metals, which in turn determines the metal distribution in the final nanoparticle. This article reports a simulation study, which analyzes how the intermicellar exchange influences reactivity, and provides a comprehensive analysis of the resulting nanostructures.

To verify the validity of the model predictions, simulation results of the arrangement of Au/Pt nanoparticles prepared in microemulsions were compared with Au/Pt nanoparticles synthesized using the same conditions of the simulation studies. The successful agreement between the theoretical and experimental STEM (scanning transmission electron microscopy) profiles confirms the validity of the simulation model [37]. The Au/Pt bimetallic system was successfully simulated by employing a reduction rate ratio of $v_{Au}/v_{Pt} \approx 10$. This value can be associated with bimetallic couples with a difference in standard reduction potential of about $\Delta\varepsilon \approx 0.2$–$0.3$ V [29]. These results can be generalized to other bimetallic nanoparticles whose $\Delta\varepsilon$ is within this range.

To design a synthesis route to obtain a bimetallic particle with a particular nanostructure, one must manipulate the metal distribution by varying the experimental conditions. It would be highly beneficial if changes in the metal arrangement could be obtained by simply changing the proportion of reactants. With the final goal of tuning the conditions for synthesizing specific bimetallic structures, we studied whether the metal distribution could be modified by a change in the proportion of reactants.

2. SIMULATION PROCEDURE

The main strategy behind the one pot method for the synthesis of Au/Pt bimetallic nanoparticles using microemulsions involves mixing three microemulsions, each containing one reactant (the two metal salts and reducing agent). After mixing, the micelles move and collide, allowing reactants to come into contact with one another due to material transfer between colliding droplets. It is assumed that this intermicellar exchange occurs when a collision between two micelles results in the merger of the micelles, establishing a water channel between them. When one of the two metal salts ($[AuCl_4]^-$ or $[PtCl_6]^{2-}$ for the preparation of Au/Pt particles) and the reducing agent (e.g., hydrazine) are located in the same micelle, the chemical reduction takes place inside the reverse micelle to obtain metal atoms (Au or Pt). That is, the droplets of the microemulsion can be conceived as tiny compartments or nano-reactors. The isolation of each nanoparticle inside a micelle avoids nanoparticle aggregation with particles located in nearby micelles. In this way, microemulsion droplets act as ideal templates, providing a compartmentalized reaction environment. The simulation model employed in this research attempts to simulate the kinetic course of the chemical reaction in these nanoreactors. In order to study the metal distribution in the final nanoparticle, the order of the metal reduction inside each micelle is stored and analyzed at the end of the synthesis.

3. REACTION MEDIA DESCRIPTION

The microemulsion structure is assumed as spherical micelles in a continuous oil phase. To reproduce this heterogeneous media, the microemulsion is defined as a set of micelles randomly located in a three dimensional lattice. Each simulation run starts with three different sets of micelles randomly distributed: micelles carrying Au salt (M-Au), Pt salt (M-Pt) and reducing agent (M-R). The volume fraction occupied by micelles is set at $\phi = 10\%$. Each micelle can act as a nanoreactor during nanoparticle synthesis. Thus, although initially each microemulsion carries only one kind of reactant, as the synthesis takes place each micelle may have species coexisting together: reactants (faster reduction metal salt $[AuCl_4]^-$, slower reduction metal salt $[PtCl_6]^{2-}$ and reducing agent R), free metal products (Au and Pt), and growing particles (aggregates composed by Au and Pt atoms).

The movement of micelles is assumed to be governed by Brownian motion. Two different strategies have been tested to simulate the movement and collisions of micelles. In the first one, micelles diffuse on a square lattice by performing random walks to contiguous lattice sites [38]. The movement obeys the exclusion principle (a lattice site can be occupied by only one micelle at the same time) and cyclic boundary conditions were implemented. A collision between micelles takes place when two micelles occupy two contiguous lattice sites. In the second method, two micelles are randomly selected to collide. Because of collision, micelles fuse to form a short-lived dimer and a water channel is established between them [39], allowing the exchange of material content. After collision, the micelles redisperse. To save computation time all collision are assumed to be effective. Similar results were obtained by using both collision models. The latter requires less computation time, thus it was implemented in this study.

4. MICROEMULSION COMPOSITION

Microemulsions consist of water nano-droplets dispersed in a continuous oil medium. The dispersion is stabilized by a monolayer of surfactant which accumulates at the oil–water interface. The flexibility of the surfactant film surrounding micelles is a parameter associated with the interfacial curvature, which depends on the interactions on both sides of the interface. The flexibility is dictated by the microemulsion composition (mainly by the surfactant), and is directly related to the facility with which intermicellar channels can be established. The intermicellar exchange of material takes place through the intermicellar channel, thus the kinetics of the nanoparticle formation will strongly depend on the channel feature.

Two aspects must be taken into account in order to establish how surfactant film flexibility is included in the simulation model. First, the material exchange between micelles will only be possible if the dimer remains intact for a sufficient amount of time. The longer that two colliding micelles stay together, the greater the number of species that can be exchanged during a collision. Therefore, dimer stability is directly related to the intermicellar exchange rate. Second, the intermicelar channel diameter restricts the size of the particles capable of crossing the channel. It is also related to the microemulsion composition, because the more flexible the film, the larger the channel size. That is, a more flexible film allows the exchange of larger particles than a rigid film. The most important factor determining the intermicellar exchange of isolated species, such as reactants and non-aggregated metallic atoms, is the dimer stability [38]. This is because free species traverse the channel one by one. Thus the channel size would not be important in this case. On the contrary, the channel size is the most important determining factor if the exchanged species is a growing particle (an aggregate of metal atoms). From this perspecive, the flexibility of the surfactant film is introduced in the model by means of two simulation parameters: the exchange parameter, k_{ex}, which dictates the exchange protocol of free species (see below), and the flexibility parameter, *f*, which restricts the size of exchanged particles. Both parameters must increase together, because a flexible film implies that a larger particle can be transferred (high *f*) and that intermicellar exchange is faster (high k_{ex}). Experimental results obtained in a rigid microemulsion, such as AOT (sodium bis(2-ethylhexyl) sulfosuccinate)/*n*-heptane/water, were succesful compared to simulation data where the flexibility was characterized by $f = 5$ and $k_{ex} = 1$ [34]. Nanoparticles obtained using a more flexible microemulsion, such as isooctane/tergitol/water, were successfully reproduced by simulation data using $f = 30$ and $k_{ex} = 5$ [37].

5. INITIAL CONCENTRATION INSIDE MICELLES

Initially, each reactant ($[AuCl_4]^-$, $[PtCl_6]^{2-}$ and R) is distributed throughout the micelles of the corresponding microemulsion according to the Poisson distribution as:

$$P(c_i) = \frac{\langle c_i \rangle^{c_i}}{c_i!} \exp\left(-\langle c_i \rangle\right).$$

where i =$[AuCl_4]^-$, $[PtCl_6]^{2-}$ or R which is the metal salt or reducing agent, c_i is the number of reactants per micelle, and $P(c_i)$ is the probability that a micelle carries c_i reactants whose average occupancy is $\langle c_i \rangle$. We present results using an average value of $\langle c \rangle$ = 64 reactants in a micelle, which corresponds to 0.40 M. Molar concentration was calculated considering the droplet radius, *r*, of a 75% isooctane/20% tergitol/5% water microemulsion (*r* = 4 nm as obtained by DLS (dynamic light scattering)). From this radius, and assuming a spherical shape ($V_{micelle} = 4/3\pi r^3$), the molar concentration of a micelle containing 64 atoms is calculated as:

$$\langle c\rangle = 64\frac{1}{V_{\mathrm{droplet}}}\frac{1}{N_{\mathrm{Av}}} = 0.40\,\mathrm{M},$$

where N_{Av} is Avogadro's number. In order to study the influence of the proportion of reactants, the average concentration is kept constant ($\langle c\rangle$ = 64 reactants per micelle) and the proportion of each reactant was varied as follows: 12.5% Au, $\langle c_{[\mathrm{AuCl_4}]^-}\rangle$ = 16, $\langle c_{[\mathrm{PtCl_6}]^{2-}}\rangle$ = 112; 25% Au, $\langle c_{[\mathrm{AuCl_4}]^-}\rangle$ = 32, $\langle c_{[\mathrm{PtCl_6}]^{2-}}\rangle$ = 96; 37.5% Au, $\langle c_{[\mathrm{AuCl_4}]^-}\rangle$ = 48, $\langle c_{[\mathrm{PtCl_6}]^{2-}}\rangle$ = 80; 50% Au, $\langle c_{[\mathrm{AuCl_4}]^-}\rangle$ = 64, $\langle c_{[\mathrm{PtCl_6}]^{2-}}\rangle$ = 64; 62.5% Au, $\langle c_{[\mathrm{AuCl_4}]^-}\rangle$ = 80, $\langle c_{[\mathrm{PtCl_6}]^{2-}}\rangle$ = 48; 75% Au, $\langle c_{[\mathrm{AuCl_4}]^-}\rangle$ = 96, $\langle c_{[\mathrm{PtCl_6}]^{2-}}\rangle$ = 32; 87.5% Au, $\langle c_{[\mathrm{AuCl_4}]^-}\rangle$ = 112, $\langle c_{[\mathrm{PtCl_6}]^{2-}}\rangle$ = 16. The reducing agent concentration $\langle c_{\mathrm{R}}\rangle$ was always double that of the average concentration of the metal precursors ($\langle c_{\mathrm{R}}\rangle$ = 128 molecules of hydrazine per micelle). For convenience, Table 1 provides a summary of the acronyms and abbreviations used throughout this article.

6. TIME UNIT BASE

The time unit is one Monte Carlo step, which is defined as follows. One Monte Carlo step starts when 10% of the micelles are chosen to collide at random. The micelles then fuse and material exchange may take place. The nature and quantity of species inside the chosen micelles can be modified according to the exchange criteria described below. Once the composition inside both micelles is updated, one Monte Carlo step (mcs) is completed. The composition inside each micelle is stored step by step, because the sequence of the metal reduction is decisive in predicting the final nanoparticle structure. These results monitor the evolution of particle distribution as a function of time. One simulation run is finished when the composition of every particle inside all micelles remains constant.

7. INTERMICELLAR EXCHANGE PROTOCOL OF REACTANTS AND METAL ATOMS

Species can be transferred between micelles during the short-lived dimer formation. This transfer is closely related to the intermicellar exchange rate, because the faster the rate, the more species that can traverse the dimer channel during a collision. The exchange criterium will be the concentration gradient, that is, species flow from a region (micelle) of higher concentration to one of lower concentration until the concentration becomes equal. A simulation parameter k_{ex} is included to restrict the maximum number of reactants ($[AuCl_4]^-$, $[PtCl_6]^{2-}$ or R) which can be exchanged between micelles during a collision. If the higher occupied micelle contains a number of molecules greater than k_{ex}, then the maximum number of reactant molecules that can cross the channel towards the micelles containing less reactants is k_{ex}. If the number of reactants to be exchanged is lower than k_{ex}, reactants are redistributed until the concentration inside both colliding micelles becomes equal after collision.

Table 1: Glossary of acronyms and abbreviations used in this article.

$[AuCl_4]^-$	faster metal precursor (Au salt)
$[PtCl_6]^{2-}$	slower metal precursor (Pt salt)
R	reducing agent
M-Au	micelle containing Au salt
M-Pt	micelle containing Pt salt
M-R	micelle containing reducing agent
Au	faster metal atom (reduction product of metal precursor $[AuCl_4]^-$)
Pt	slower metal atom (reduction product of metal precursor $[PtCl_6]^{2-}$)
$P(c_i)$	probability that a micelle carries c_i reactants ($[AuCl_4]^-$, $[PtCl_6]^{2-}$ or R)
$\langle c_{[AuCl_4]^-} \rangle$	mean number of metal precursor $[AuCl_4]^-$ at the beginning
$\langle c_{[PtCl_6]^{2-}} \rangle$	mean number of metal precursor $[PtCl_6]^{2-}$ at the beginning
$\langle c_R \rangle$	mean number of reduced molecules R at the beginning
k_{ex}	exchange parameter: number of free units ($[AuCl_4]^-$, $[PtCl_6]^{2-}$, Au, Pt, R) which could be transferred during a collision
v_{Au}	faster chemical reduction rate: percentage of $[AuCl_4]^-$ inside the colliding droplets which gives rise to products
v_{Pt}	slower chemical reduction rate: percentage of $[PtCl_6]^{2-}$ inside the colliding droplets which gives rise to products
n_{Au}	number of Au atoms (reduction product of faster metal precursor $[AuCl_4]^-$)
n_{Pt}	number of Pt atoms (reduction product of slower metal precursor $[PtCl_6]^{2-}$)
dn_{Au}/dt	Au reaction rate; calculated from simulation data
dn_{Pt}/dt	Pt reaction rate; calculated from simulation data
n_A^*	critical nucleus size: minimum number of A atoms required to form a stable nucleus
n_B^*	critical nucleus size: minimum number of B atoms required to form a stable nucleus
$n_{A\text{-}B}^*$	heterogeneous critical size: minimum number of A and B required to form a stable nucleus
f	film flexibility: maximum particle size for transfer between droplets
q	maximum number of metal atoms which can be carried by a droplet

It is assumed that k_{ex} is mainly determined by the microemulsion composition [39-42] and the nature of the material is less important. Thus, although the characteristics of the species traversing the channel could modify the intermicellar exchange rate, a single value of k_{ex} was used in this investigation ($k_{ex,[AuCl_4]^-} = k_{ex,[PtCl_6]^{2-}} = k_{ex,R} = k_{ex}$).

8. CHEMICAL REDUCTION RATES

Due to the redistribution of material between the micelles, one metal salt and the reducer can be located inside the same micelle in order for chemical reduction to take place. The reduction potentials of the two metal salts, $[AuCl_4]^-$ and $[PtCl_6]^{2-}$, are different, so both chemical reductions will occur at different rates. The two metal salts are reduced according to the simulation parameter, v_i, where *i* corresponds to the metal salt, $[AuCl_4]^-$ or $[PtCl_6]^{2-}$. This parameter represents the chemical reduction rate as the probability of obtaining one reduced metal atom from each metal salt/reducing agent molecule pair available in the micelle. Because of the faster reduction rate of Au, it is assigned a value of $v_{Au} = 1$, that is, 100% of reactants inside micelles gives rise to products. This implies that the reaction is completed and only Au atoms and excess reactants (either $[AuCl_4]^-$ or R) are to be distributed to daughter micelles. Hence, the reactants $[AuCl_4]^-$ and R will not coexist in a micelle. In order to consider different reactions rates, the probability of reactants located in the micelle that reduces to metal atoms can be decreased. In this study we present results where this parameter is fixed at $v_{Au}/v_{Pt} = 10$. This means that only one of every ten pairs of $[PtCl_6]^{2-}$ and a reducing agent available in the micelle produces Pt atoms. The remainder of the $[PtCl_6]^{2-}$ and reducing agent that did not react remains in the micelle and can be exchanged or can react in a future collision. When both reduction reactions are possible because all three reactants ($[AuCl_4]^-$, $[PtCl_6]^{2-}$ and R) are located inside the same micelle at the same time, both reductions are allowed to take place during the same collision.

9. NUCLEATION

Nucleation is the process by which atoms (ions or molecules), which were initially isolated in solution, arrange to form a thermodynamically stable nucleus. The nucleus grows as more atoms are deposited onto it. A new phase (nucleus) begins to form from random fluctuations. Because a nucleation event is hindered by energy barriers, only fluctuations which overcome these nucleation barriers can give rise to the formation of a nucleus capable of further growth [43]. Classical nucleation theory establishes the existence of a critical nucleus size from which a nucleus will grow instead of dissolving. A nucleus smaller than this critical size will spontaneously dissolve. In microemulsions, nucleation similarly requires the presence of enough atoms to exceed the critical nucleation size inside the same micelle [44]. Nucleation is included in the model by means of the variable critical nucleus, n^*, which is compared with the actual amount of metal atoms inside the same micelle. If it is smaller than n^*, the atoms remain free (non-aggregated) inside the micelle because the nucleus is considered to be unstable and will disintegrate to produce isolated atoms. This implies that atoms can be exchanged during a subsequent collision subject to k_{ex} parameter. However, if the number of reduced atoms located inside the same micelle exceeds n^*, all atoms gather producing a stable nucleus capable of further growth. The exchange of nucleii between colliding micelles depends on the intermicellar channel size, since an aggregate of atoms must be exchanged as a whole. This type of material exchange is governed by the film flexibility parameter, *f* (see below).

The total Gibbs energy of an atomic arrangement determines the alloying ability of a bimetallic nanoparticle A-B. It depends on composition because of the different atomic binding energy of A-A, B-B, and A-B species. These different binding energies (A-A, B-B, and A-B) dictate the minimum size that must be reached by atoms for nucleation depending on the composition. This phenomena is included in the simulation model by means of the inclusion of three critical nucleus numbers (n_A^*, n_B^* and $n_{A\text{-}B}^*$). Once the critical number is exceeded, the cluster grows by deposition of all metal atoms (either Au or Pt) located inside the same micelle. It is assumed that only one particle can be carried by a micelle. That is, only one nucleation event is possible in each micelle.

The influence of the critical size on the formation of bimetallic nanoparticles in micelles was previously studied[36,45]. In this paper, n^*

was kept constant ($n_A^* = n_B^* = n_{A-B}^* = 1$) so as not to interfere in the discussion.

10. NANOPARTICLE GROWTH

It is assumed that nanoparticles grow by deposition of metallic atoms on the nucleus, that is, a growing particle is built up layer by layer. Therefore, whenever a metal atom is exchanged towards a nucleated micelle, this atom is deposited on the nucleus. The sequence of metal deposition (Au or Pt) is stored in each micelle at each step.

In addition, nanoparticles can grow by autocatalysis [23,34,46]. This type of growth appears at advanced stages during the synthesis when collision between micelles containing reactants and a growing particle becomes frequent. An autocatalytic reduction is simulated by introducing two new criteria. When one of the colliding micelles is carrying a particle, the reduction will occur at double the rate and always at the existing particle. When both colliding micelles are nucleated, reduction takes place in the micelle containing the larger particle, and the atom is deposited onto it. These criteria allow us to include in the model the well-known belief that a bigger surface (bigger particle) has a greater probability to act as a catalyst. In this way, the growth of a preexistent nucleus is favored instead of the formation of a new one. To simplify, autocatalytic growth is governed by particle size, without taking into account the surface composition (Au or Pt).

Another type of growth is Ostwald ripening, that is, the growth of larger particles at the expense of smaller ones by transport of material. It is assumed that the easier solubilization of the smallest particles causes their decrease in size. These atoms/molecules, free in solution, will deposit on the largest particles. That is, large particles grow even larger, drawing material from the smaller ones, which shrink. This possibility is included in the exchange protocol of particles described as follows.

11. INTERMICELLAR EXCHANGE PROTOCOL OF GROWING PARTICLES

The transfer of a growing particle to another micelle also containing a particle is possible whenever the comunication channel of the colliding micelles is sufficiently large. As mentioned before, the size of the channel is represented by the f parameter, which is associated with surfactant film flexibility. A collision between micelles that both contain a growing particle can result in a unidirectional transfer of material. Thus, the smaller particle always moves from the initial micelle to the micelle containg the larger particle. In this way, two particles in two colliding micelles give rise to a single particle. This exchange is only possible when the smaller particle contains less than $f$atoms (i.e., it can traverse intermicellar channel). This rule is defined simply by the particle size, thus the composition (Au or Pt) is not taken into consideration. All types of material exchange (reactants, isolated atoms and growing particles) are allowed to be exchanged during the same collision.

12. MICELLE SIZE

The nanoparticle growth may be restricted by the size of the micelles because the surfactant film covering the micelle has a finite bending modulus. The simulation includes a micelle size parameter, which limits the nanoparticle size, establishing a maximum quantity of metal products which can be located inside the same micelle. In this study we present results using low values of reactant concentrations, thus the influence of micelle size on nanoparticle growth is assumed to be negligible.

13. METAL DISTRIBUTION IN BIMETALLIC NANOPARTICLES

Nanoparticle formation is complete when the contents of each micelle do not vary with time. Each simulation run results in a set of micelles, each of which can contain a particle whose composition can be different. The composition of each particle is stored during the process. The final metal distribution in the nanoparticle is determined by the order in which the two metals are deposited onto the nanoparticle surface. Given the assumption that the particles are spherical, clustering steps give rise to the build up of concentric layers by adding new atoms over a particle. For each nanoparticle, the sequence of metals is stored and divided into ten concentric layers. The number of particles which contain a given percentage of Au is monitored from the inner layer (core) to the outer layer (shell), thus distribution and average composition (percent Au) can be calculated layer by layer. This final distribution is averaged over 1000 runs.

The nanoparticle structure is represented by histograms, in which the layer composition (% Au) is represented by a color grading. The distribution ranges from blue (0–10% for the fast metal Au) to red (90–100% of Au). 50% of each metal is represented by grey. Lighter colors represent a higher proportion of pure metal in the layer. The histograms show how many particles contain a given percentage of Au in each layer. In this way, the histograms allow analysis of the variation of metal arrangement from the early stages (core) as the synthesis reaches the shell formation. The nanoparticle structure is also represented by means of concentric spheres whose thickness is proportional to the number of layers for a given composition, keeping the same color scheme.

14. RESULTS AND DISCUSSION

14.1 Metal distribution in Au–Pt nanoparticles for different Au:Pt ratios

A set of experiments were carried out varying the proportion between Au and Pt salts. Due to the complex mechanism governing the reactions inside the micelles, most of the parameters have been kept fixed in order not to interfere with the computational study. The following parameters are fixed: average metal concentration inside the droplets $\langle c \rangle$ = 64 metal precursors per micelle, microemulsion composition (characterized by the exchange parameter, k_{ex} = 1 and the film flexibility parameter f = 30), and critical nucleus sizes $n_A^* = n_B^* = n_{A\text{-}B}^* = 1$. The distribution of both metals within a nanoparticle can be observed in Figure 1, which shows the structures obtained by the simulation. When the quantities of the metal salts are equal (see Figure 1a, 50% Au), the nanoparticle shows a core–shell structure due to the difference in reduction rates ($v_{Au} = 10 \cdot v_{Pt}$). Most particles have a core composed by the faster reduction metal Au (see red bars on the left), followed by mixed middle layers, and then outer layers composed by Pt forming the shell (see blue bars behind on the right). As the Au proportion is larger (see the top row of Figure 1), the amount of Pt in the shell progressively diminishes until it disappears. Thus, 62.5% Au gives rise to a particle with a Pt-enriched shell (see Figure 1b), 75% Au gives rise to a mixed outer layer (see Figure 1c), and 87.5% Au particle does not result in a shell at all. Although Pt is mainly located in the outer layers, a few particles whose statistical weight is negligible result in a Pt core (see blue bars in front on the right).

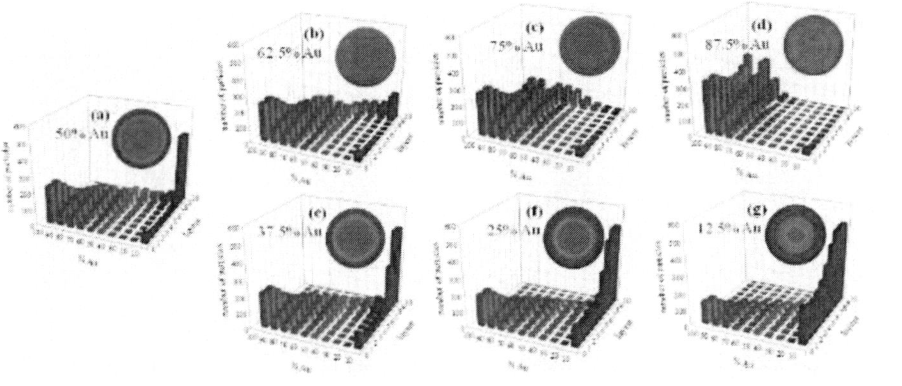

Figure 1: Histograms reprensent the number of particles with a given percentage of Au in each layer, from the nanoparticle core to the surface. Synthesis conditions: reduction rate ratio (v_{Au}/v_{Pt} = 10), film flexibility (k_{ex} = 5, f = 30) and average reactant concentration ($\langle c \rangle$ = 64 metal ions in a micelle). Each histogram shows a different proportion between metals salts: a) 50% Au, $\langle c_{[AuCl_4]^-} \rangle$ = 64, $\langle c_{[PtCl_6]^{2-}} \rangle$ = 64; b) 62.5% Au, $\langle c_{[AuCl_4]^-} \rangle$ = 80, $\langle c_{[PtCl_6]^{2-}} \rangle$ = 48; c) 75% Au, $\langle c_{[AuCl_4]^-} \rangle$ = 96, $\langle c_{[PtCl_6]^{2-}} \rangle$ = 32; d) 87.5% Au, $\langle c_{[AuCl_4]^-} \rangle$ = 112, $\langle c_{[PtCl_6]^{2-}} \rangle$ = 16; e) 37.5% Au, $\langle c_{[AuCl_4]^-} \rangle$ = 48, $\langle c_{[PtCl_6]^{2-}} \rangle$ = 80; f) 25% Au, $\langle c_{[AuCl_4]^-} \rangle$ = 32, $\langle c_{[PtCl_6]^{2-}} \rangle$ = 96; g) 12.5% Au, $\langle c_{[AuCl_4]^-} \rangle$ = 16, $\langle c_{[PtCl_6]^{2-}} \rangle$ = 112 metal ions in a micelle. Scheme color: blue (0–45% of Au), grey (45–55% of Au), red (55–100% of Au). Brighter red corresponds to more Au. The circles in each histogram represent the nanoparticle structure in concentric layers, keeping the same color scheme.

With the decrease of Au proportion (see Figure 1e, f and g), the number of outer layers with pure Pt composition becomes higher, that is, the thickness of the Pt shell increases. This is an expected result, because if Pt reduces later, the resulting Pt atoms will deposit forming the more external layers. The higher the amount of Pt salt, the thicker the Pt shell. Regarding the cores, it is interesting to point out that the number of particles with a Pt core increases with the increase of % Pt (see blue bars in front on the right in bottom file). That is, when Pt is abundant (see Figure 1g, 12.5% Au) the number of particles with a pure Pt core is greater than the number of particles with a pure Au core. This is an unexpected result because it implies that an important amount of Pt is reduced at the very beginning of the synthesis, despite its slower reduction rate. It means that Au–Pt nanoparticles with a Pt-core can be obtained. Similar qualitative behavior is obtained when different average concentrations are simulated. This unpredicted evolution of metal segregation when the proportions are varied can be understood by means of a deeper kinetic study.

14.2 Kinetic study

In micellar media, the diffusion of the reactants can be largely retarded or prevented [47]. Our hypothesis is that the resulting nanostructure is due to the particular combination of three main factors: the reduction rate ratio between both metals, the intermicellar exchange rate and the amount of each metal precursor inside micelles. These factors together will determine the particular sequence of deposition of the metals, which in turn will determine the metal distribution in the final nanoparticle.

As the synthesis advances, the simulation monitors the number and type of each atom as a function of time.Figure 2 shows the number of Au and Pt atoms produced in all micelles as the reaction proceeds. This figure corresponds to the same synthesis conditions shown in Figure 1. The continuous lines in Figure 2 represent the number of Au atoms produced at different initial Au proportions and discontinuous lines correspond to Pt atoms. Each Au curve (continuous line) has a corresponding Pt curve (discontinuous line in the same color): for example, an initial concentration $\langle c_{[AuCl_4]^-} \rangle$ = 96, $\langle c_{[PtCl_6]^{2-}} \rangle$ = 32 is represented as a medium shade of red (75% Au, continuous line, and 25% Pt, discontinuous line). It can be clearly observed that the Au (which is reduced faster) is obtained earlier, even when the Au is the minority metal (compare the 12.5% Au and the 87.5% Pt lines). All curves lead to a plateau when the reactants have been exhausted. As initial metal salt amount is increased this plateau is reached at later stages of the synthesis, as expected. One interesting observation from these results is that the slopes of all the Au curves are similar. On the contrary, the Pt curves are strongly dependent on the initial Pt quantity. This implies that the influence of metal quantity on the actual metal production rate is different for Au than for Pt, as will be discussed later. Similar qualitative behavior was found using different average concentration values.

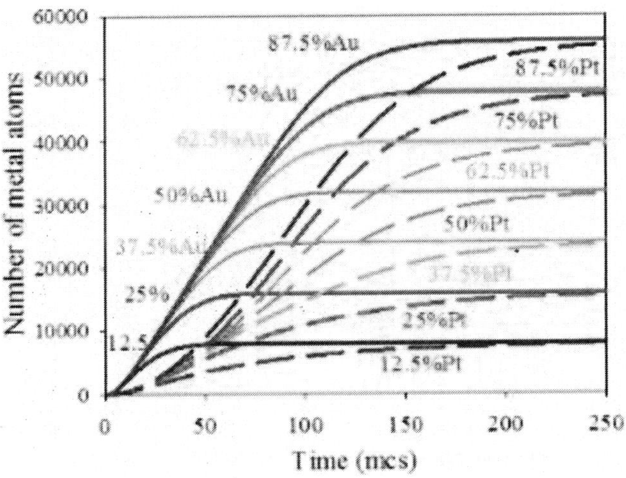

Figure 2: Time evolution of the number of metal atoms obtained in micelles. Continuous and discontinuous lines show the resulting Au and Pt, respectively. Synthesis conditions: flexible film (k_{ex} = 5, f = 30), reduction rate ratio v_{Au}/v_{Pt} = 10; average concentration $\langle c \rangle$ = 64. Scheme color: dark blue lines: 12.5% Au, $\langle c_{[AuCl_4]^-} \rangle$ = 16, $\langle c_{[PtCl_6]^{2-}} \rangle$ = 112; middle blue lines: 25% Au, $\langle c_{[AuCl_4]^-} \rangle$ = 32, $\langle c_{[PtCl_6]^{2-}} \rangle$ = 96; light blue lines: 37.5% Au, $\langle c_{[AuCl_4]^-} \rangle$ = 48, $\langle c_{[PtCl_6]^{2-}} \rangle$ = 80; grey lines: 50% Au, $\langle c_{[AuCl_4]^-} \rangle$ = 64, $\langle c_{[PtCl_6]^{2-}} \rangle$ = 64; light red lines: 62.5% Au, $\langle c_{[AuCl_4]^-} \rangle$ = 80, $\langle c_{[PtCl_6]^{2-}} \rangle$ = 48; middle red lines: 75% Au, $\langle c_{[AuCl_4]^-} \rangle$ = 96, $\langle c_{[PtCl_6]^{2-}} \rangle$ = 32; dark red lines: 87.5% Au, $\langle c_{[AuCl_4]^-} \rangle$ = 112, $\langle c_{[PtCl_6]^{2-}} \rangle$ = 16.

The first aspect to discuss is how to relate the resulting metal curves to the nanoparticle structure. For a better understanding of the metal segregation in a nanoparticle, Figure 3 shows 25%, 50% and 75% Au curves with the corresponding Pt curves (the same as Figure 2). The dotted black line shows the difference between the number of Au and Pt atoms (n_{Au} – n_{Pt}) as a function of time for each Au proportion. This difference increases as time goes on until a maximum is reached when the Au salt is almost exhausted. From this point, if Pt salt is also exhausted (see Figure 3a, 75% Au) due to the small amount of Pt, the n_{Au} – n_{Pt} curve remains aproximately constant. If the quantity of Pt salt is larger (see Figure 3b, 50% Au), the Pt continues to produce when the maximum is reached, and the difference diminishes until it becomes insignificant when slow Pt reduction finishes. Finally, if the minority metal is Au (see Figure 3c, 25% Au), thus Au finishes earlier, while Pt reduction continues. Consequently, the maximum is also reached earlier, and the difference n_{Au} – n_{Pt} takes negative values as times goes on. As the Au proportion diminishes, the maximum is not only displaced towards earlier stages but also the magnitude diminishes. This maximum reflects the largest difference in metal product availability, therefore its temporal localization and its magnitude will be directly correlated to the nanoparticle structure. Therefore, if a quite pronounced maximum appears at the beginning of the synthesis, a large quantity of Au atoms are obtained when the core is forming, and the core will be essentially homogeneous and composed of Au. From the maximum, Pt and the remaining Au are deposited on the core, giving rise to the middle mixed layers. In this case, a pure Pt shell is formed when the Au is exhausted. This is the case shown in Figure 3b, which corresponds to the core–shell structure obtained using 50% Au. A high maximum located at later stages as shown in Figure 3a (75% Au) implies the formation of many Au-composed layers. In this case, the fact that curve does not decay after the maximum is reflected in the mixed shell, that is, there is not enough Pt in the reaction media to build up an enriched Pt shell (see also Figure 1c). Finally, the short and early maximum shown in Figure 3c (25% Au) gives rise to a particle in which the small quantity of Au is mainly located in the inner layers, as showed in Figure 1f.

A second aspect to study is understanding why the Pt reduction rate is more affected by an increase in reactant quantity than the Au rate. This implies that the influence of metal amount on the actual metal production rates is different in Au than in Pt. To gain more insight into how confinement influences chemical reactivity, the reaction rate of each metal was calculated from the slopes of the curves shown in Figure 2 as time progresses. In this way, both contributions (intermicellar exhange rate and chemical reduction rate) are taken into account to determine the reaction rate (dn_{metal}/dt). To explain the results clearly, the chemical reduction rates (deduced from the standard potentials as v_{Au}/v_{Pt} = 10) must be distinguished from the reaction rates (calculated from simulation data as

dn_{metal}/dt). Continuous and discontinuous lines in Figure 4 represent the resulting Au and Pt reaction rates, respectively, at different % Au, as the synthesis advances. The most impressive result in this figure is the profile of the rates, that is, an usual decay versus time is preceded by an increasing rate until a maximum is reached. To understand this result, one must keep in mind that the first requirement, before chemical reduction is possible, is the localization of the two reactants inside the same micelle.This implies an intermicellar exchange of material, whose rate is determined by microemulsion composition. The rate of the chemical reduction inside a micelle not only depends on the reduction rate but also on the rate of the intermicellar exchange. A priori, one could think that the intermicellar exchange rate equally affects both chemical reductions. But the interplay between exchange rate and reduction rate depends on the particular metal, as shown in Figure 4. In all cases it can be observed that the greater the metal amount, the quicker the actual reduction rate, as expected. However, the behavior of Au and Pt as increasing the corresponding metal amount is very different. Before the maximum, the Pt reaches a progressively faster rate as the Pt amount increases. On the contrary, the slope of Au rate does not depend on % Au and seems to reach a threshold from which it cannot increase anymore. In relation to the decay, Pt and Au also display different behavior. The decay slopes are determined by % Pt, but remain almost constant in the case of Au.

Firstly, we will discuss the Au reaction rate (see continuous lines in Figure 4). Previous results suggested that the compartmentalization of reaction media causes the Au reduction to be mainly controlled by the intermicellar exchange rate. The reason is that if reduction is very fast (as the case of Au), the slower step is the material intermicellar exchange, which allows for the reactants to interact. At this point, all Au precursors confined in the same micelle are reduced rapidly. In the simulation model, the intermicellar exchange rate is simulated by the parameter k_{ex}, which quantifies how many ions/molecules can be transferred between colliding micelles. In this study, we present results obtained by allowing a maximum intermicellar exchange of five reactants (Au salt, Pt salt and/or reducing agent). That is, independent of the Au amount inside the micelle, only $k_{ex} = 5$ reactants are allowed to be exchanged in each collision. As a result, although an instantaneous reaction is simulated for Au reduction, only a maximum of five Au atoms can be obtained during each effective collision. This restriction allows us to understand the Au rate profile. During the first steps (which determine the core formation), the reaction rate increases until it reaches the intermicellar control. Then, the Au reaction rate decreases as the Au salts are finished. Some time is needed to reach the intermicellar control. At the beginning of the synthesis, microemulsions containing the reactants are mixed. Due to Brownian motion, micelles collide with each other, but only collisions between one micelle containing Au salt (M-Au) and another micelle containing the reduction agent (M-R) allow both reactants to be found inside the same waterpool, where the reduction then takes place. The rest of the possible collisions (M-Au and M-Au, M-Au and M-Pt, M-Pt and M-Pt, M-Pt and M-R, M-R and M-R) can not give rise to Au atoms. These types of collisions simply redistribute reactants between micelles. As the reactants are redistributed between micelles, more collisions will be effective, providing metallic atoms and an increasing reaction rate. The speed at which the intermicellar control is achieved does not depend on the Au amount, but only on the material intermicellar exchange rate. As a consequence, all slopes at the beginning of the Au curves in Figure 4 are equal. If the Au amount is scarce, reaction rate does not reach intermicellar control due to the Au salt finishing earlier.

Larger values of the reaction rate are reached by increasing the Au salt quantity, as expected. In addition, when the Au reaction rate achieves the exchange rate, it remains constant even though the Au amount inside the micelles would be larger (compare the curves at higher % Au). Likewise, according to our results, the decrease in rate is also not influenced by the Au amount, because once the Au salt is redistributed, the Au reduction only depends on the intermicellar exchange rate, that is, it steadily diminishes while there is Au precursor present. To summarize, due to the intermicellar exchange restrictions, the Au reaction rate is controlled by the material intermicellar exchange. Similar qualitative behavior was obtained using different total concentrations, as shown inFigure 5. The combination of a low average concentration and a low Au proportion leads to an initial Au quantity smaller than k_{ex}, (see Figure 5a, $\langle c \rangle$ = 32, dark blue continuous line: 12.5% Au, $\langle c_{[AuCl_4]^-} \rangle$ = 4, $\langle c_{[PtCl_6]^{2-}} \rangle$ = 28), and the initial slope does not reach the exchange control and the reactant concentration appears to be the controlling factor, as expected. Likewise, it can be observed that for a given proportion of reactants, faster reaction rates are reached at a higher concentration (compare Figure 5 a, b and c at a fixed % Au). In addition, as concentration increases, the plateau is reached at lower % Au and remains longer (see Figure 5c, $\langle c \rangle$ = 128, at higher % Au). Finally, it is interesting to emphasize that if synthesis conditions lead to the plateau achievement, the classical belief that the larger the reactant amount, the faster the reaction rate is not valid in this confined media.

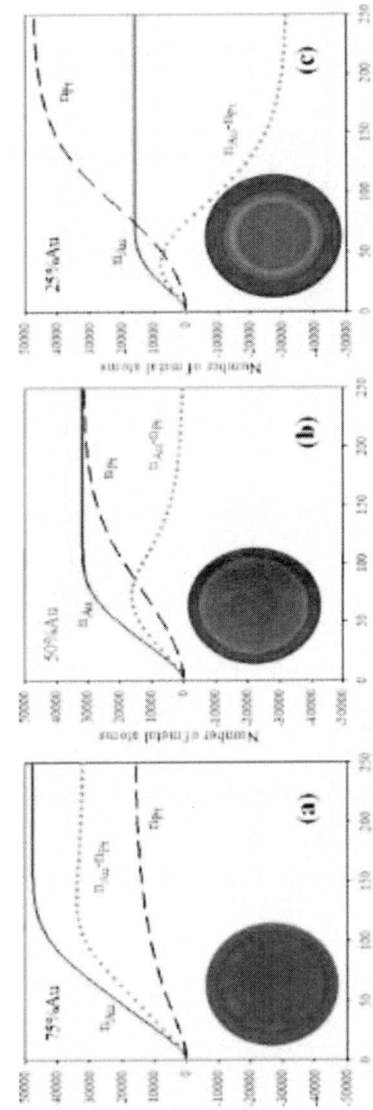

Figure 3: Time evolution of the number of metal atoms obtained in micelles. Continuous and discontinuous lines show the resulting Au and Pt, respectively. Dotted lines represent the difference between the number of Au and Pt atoms. Synthesis conditions: flexible film (k_{ex} = 5, f = 30), reduction rate ratio v_{Au}/v_{Pt} = 10, average concentration $\langle c \rangle$ = 64. a) 75% Au, $\langle c_{[AuCl_4]^-} \rangle$ = 96, $\langle c_{[PtCl_6]^{2-}} \rangle$ = 32; b) 50% Au, $\langle c_{[AuCl_4]^-} \rangle$ = 64, $\langle c_{[PtCl_6]^{2-}} \rangle$ = 64; c) 25% Au, $\langle c_{[AuCl_4]^-} \rangle$ = 32, $\langle c_{[PtCl_6]^{2-}} \rangle$ = 96.

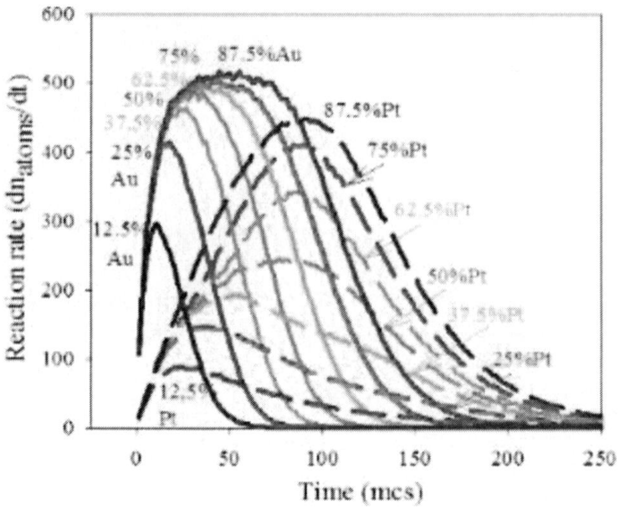

Figure 4: Reaction rate versus time. Continuous and discontinuous lines show the resulting Au and Pt, respectively. Synthesis conditions: flexible film (k_{ex} = 5, f = 30), reduction rate ratio v_{Au}/v_{Pt} = 10; average concentration $\langle c \rangle$ = 64. Scheme color: dark blue lines: 12.5% Au, $\langle c_{[AuCl_4]^-} \rangle$ = 16, $\langle c_{[PtCl_6]^{2-}} \rangle$ = 112; medium blue lines: 25% Au, $\langle c_{[AuCl_4]^-} \rangle$ = 32, $\langle c_{[PtCl_6]^{2-}} \rangle$ = 96; light blue lines: 37.5% Au, $\langle c_{[AuCl_4]^-} \rangle$ = 48, $\langle c_{[PtCl_6]^{2-}} \rangle$ = 80; grey lines: 50% Au, $\langle c_{[AuCl_4]^-} \rangle$ = 64, $\langle c_{[PtCl_6]^{2-}} \rangle$ = 64; light red lines: 62.5% Au, $\langle c_{[AuCl_4]^-} \rangle$ = 80, $\langle c_{[PtCl_6]^{2-}} \rangle$ = 48; medium red lines: 75% Au, $\langle c_{[AuCl_4]^-} \rangle$ = 96, $\langle c_{[PtCl_6]^{2-}} \rangle$ = 32; dark red lines: 87.5% Au, $\langle c_{[AuCl_4]^-} \rangle$ = 112, $\langle c_{[PtCl_6]^{2-}} \rangle$ = 16 reactants per micelle.

Secondly, the reaction rate of the Pt reduction is strongly dependent on the Pt amount. Discontinuous lines inFigure 4 show that the higher the Pt proportion, the larger the reaction rate of Pt during all the synthesis. The rate of intermicellar exchange is the same for both metals, so the reason for this different behavior is not obvious. Previous studies suggest that micelles may act as a cage [27,48]. This assumption is based on the well-known cage effect in chemical kinetics. It is assumed that encounters (the process in which both reactants diffuse together to become neighbors) take place in a different way in a solution than in a gas. In a solution, the reactant molecules will generally jump from hole to hole in the solvent matrix. Occasionally, reactants will find themselves in the same solvent cage, so the two reactants can eventually form an encounter pair. The encounter pair can fail to react the first time, but it has many more opportunities as long as it remains in the same solvent cage. On the contrary, if the reaction takes place in the gas phase, the molecules have the freedom to go anywhere, thus many collisions will be encountered in pairs. However, if the collision should fail to be energetically or geometrically viable, the reactant molecules move away and are unlikely to meet again soon. Our hypothesis is that this approach can be used to compare the chemical kinetics in micelles and in a solution, instead of in liquid and in gas phases. That is, micelles allow for less encounters, but the reactants stay near each other for much longer than in a solution (in which there would be more encounters, but shorter time together). This assumption provides an explanation for the dependence of the Pt reaction rate on the reactant proportion. Although the intermicellar exchange rate is the same for both reactants, the Pt reduction rate is slower than the Au reduction rate. Only 10% of reactants (Pt salt and reducing agent) carried by the same micelle react to produce Pt atoms in each collision. The remaining 90% that did not react, stay in the micelle, giving rise to a reactant accumulation. The implications of this increase in concentration inside the micelles determine the kinetics of the Pt reduction. Although only a 10% of the reactants react, the amount of pairs of reactants inside the same micelle which are available to react is much higher than the pairs exchanged during the last collision. So the Pt reaction rate not only depends on the exchange constant (as the case of Au) but also on the reactant accumulation. As a result, the Pt reactant confinement will strongly affect the Pt reduction rate. This is because while there is enough reactant inside a micelle, the Pt reduction proceeds without depending on a new intermicellar exchange. This results in an increase of the Pt reduction rate due to the cage-like effect. This cage-like effect does not concern the Au reduction, because Au precursors are not accumulated due to an instantaneous reduction. On this basis, the increasing slopes before the peak of the discontinuous lines in Figure 4 can be explained by taking into account that more reactants (higher % Pt) lead to more accumulation of Pt, so the cage-like effect becomes more pronounced as % Pt increases. Therefore, if the Pt precursor amount is large enough, Pt curves must reach the intermicellar control. This behavior can be observed in the blue discontinuous line (87.5% Pt) in Figure 5c, which shows results using a larger average concentration ($\langle c \rangle$ = 128 reactants per micelle).

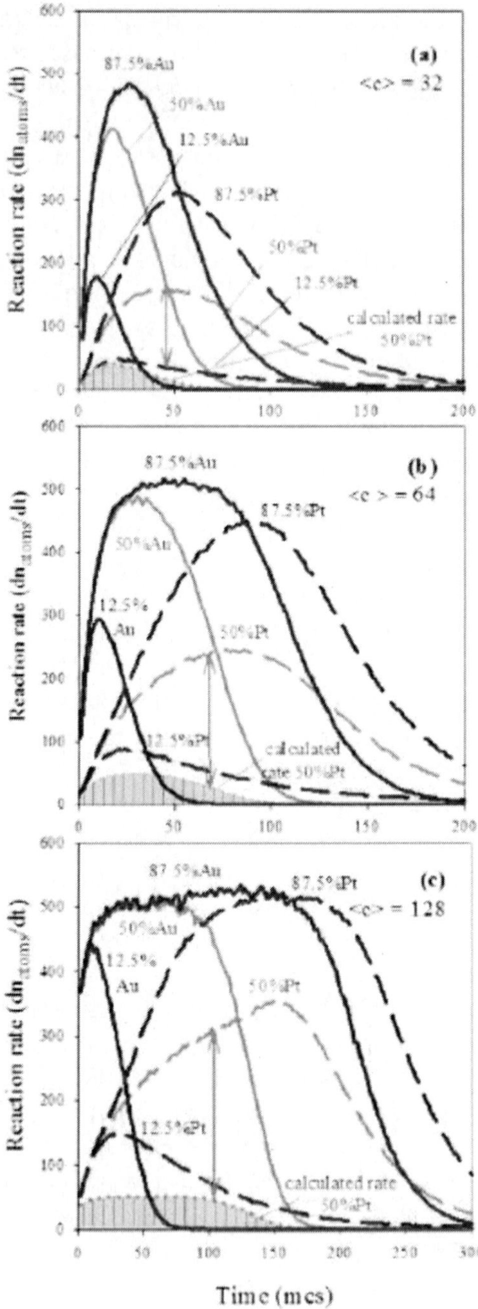

Figure 5: Reaction rate versus time. Continuous and discontinuous lines show the resulting Au and Pt, respectively. Synthesis conditions: flexible film (k_{ex} = 5, f = 30), reduction rate ratio v_{Au}/v_{Pt} = 10. Scheme color: dark

blue lines: 12.5% Au, grey lines: 50% Au, dark red lines: 87.5% Au. Curves delimiting grey areas shows the calculated Pt rate as 0.10 times the Au reaction rate: $v_{Pt,calculated} = 0.10 \cdot dn_{Au}/dt$. a) average concentration $\langle c \rangle$ = 32 reactants per micelle (12.5% Au, $\langle c_{[AuCl_4]^-} \rangle$ = 8, $\langle c_{[PtCl_6]^{2-}} \rangle$ = 56; 50% Au, $\langle c_{[AuCl_4]^-} \rangle = \langle c_{[PtCl_6]^{2-}} \rangle$ = 32; 87.5% Au, $\langle c_{[AuCl_4]^-} \rangle$ = 56, $\langle c_{[PtCl_6]^{2-}} \rangle$ = 8); b) average concentration $\langle c \rangle$ = 64 reactants per micelle (12.5% Au, $\langle c_{[AuCl_4]^-} \rangle$ = 16, $\langle c_{[PtCl_6]^{2-}} \rangle$ = 112; 50% Au, $\langle c_{[AuCl_4]^-} \rangle = \langle c_{[PtCl_6]^{2-}} \rangle$ = 64; 87.5% Au, $\langle c_{[AuCl_4]^-} \rangle$ = 112, $\langle c_{[PtCl_6]^{2-}} \rangle$ = 16); c) average concentration $\langle c \rangle$ = 128 reactants per micelle (12.5% Au, $\langle c_{[AuCl_4]^-} \rangle$ = 32, $\langle c_{[PtCl_6]^{2-}} \rangle$ = 224; 50% Au, $\langle c_{[AuCl_4]^-} \rangle = \langle c_{[PtCl_6]^{2-}} \rangle$ = 128; 87.5% Au, $\langle c_{[AuCl_4]^-} \rangle$ = 224, $\langle c_{[PtCl_6]^{2-}} \rangle$ = 32).

After the peaks in Figure 4, the fact that the slopes of the Pt reactions rates strongly depend on % Pt (in contrast with Au) is also accounted for by the cage effect. At later stages of the synthesis, local accumulation of reactants inside the micelles diminishes because reactants have been finished as reaction proceeds. So the Pt reaction rate continuously decays, and the rate of this decay will be faster as the reactants are consumed faster. Then, if the cage-like effect takes place to a greater extent at higher % Pt, it also results in a quicker decay of Pt reaction rate.

Further evidence of the acceleration of Pt reduction in micelles can be achieved if the rate of Pt reduction is theoretically calculated for the case of the cage-like effect not taking place. By assuming that the Au reduction is as quick as the intermicellar exchange allows, the Au reaction rate (dn_{Au}/dt) showed in Figure 4 will reflect the intermicellar dynamics, which does not depend on the kind of exchanged metal salt. If the Pt reduction is ten times slower than that of Au, and no Pt precursors accumulate inside micelles, then theoretically, the Pt reaction rate should be ten times slower than the Au reaction rate. A theoretical Pt reaction rate, calculated as $v_{Pt,calculated} = 0.10 \cdot dn_{Au}/dt$, would take into account the restrictions due to the intermicellar rate (reflected in dn_{Au}/dt) without considering reactant accumulation. This theoretical Pt reaction rate is shown as the curve delimiting the grey filled areas in Figure 5. This estimated Pt rate can only be calculated by equal proportion of Au and Pt salts (50% Au), so that the initial quantity of metal salt does not interfere in discussion. Lines with double arrow are drawn to associate the theoretical Pt rate with the Pt reaction rate obtained from simulation data. Our assumption is that the large difference between both curves is due to the cage-like effect. The resulting gap is larger as concentration is higher, as is expected.

Summarizing, reactant confinement affects slow and fast reducting metals in different ways. The reaction rate of Au (faster metal) is controlled by the intermicellar exchange rate, so that once the exchange control is reached, the Au rate does not vary with the Au salt proportion. However, the reaction rate of Pt (slower metal) is strongly dependent on the Pt salt proportion during the entire synthesis due to the cage-like effect.

14.3 Chemical kinetics and metal arrangement

The interplay between the reduction rates of both metals and the compartmentalization of the reaction media have key repercussions on the sequence of metal reduction. The resulting difference in reaction rates enables us to explain the modifications in metal segregation.

Due to the difference in the reduction rates, 50% Au gives rise to the typical core–shell structure. As % Au is increased, the quantity of Pt available to form the shell diminishes, and the Pt-enrichment in the shell progresively disappears.

When the % Pt is increased, the Pt reduction rate increases because of higher Pt quantity and the resulting cage-like effect. Consequently, more Pt is reduced sooner forming the inner layers, as clearly observed in Figure 1. More particles have a core composed of Pt as the % Pt is higher (see increasing blue bars in the inner layers inFigure 1, from 50 to 125% Au).

It is important to highlight that Au primarily comprises the core, even at the lowest Au percentages. This means that the acceleration of Pt reduction (due to a higher Pt quantity and the cage-like effect) cannot overtake the quicker Au reduction rate, even when a high % Pt is used.

A high % Pt gives rise to the formation of more external layers composed of Pt, so the thickness of the Pt shell increases with % Pt. This is a consequence of the excess of Pt, because once all of the Au salt is reduced forming the core, the nanoparticle can only grow by deposition of Pt.

15. CONCLUSION

Intermicellar exchange is a factor of critical importance in metal reduction reactions in micelles, because its impact is different depending on the reduction rates of each metal. If the reduction rate is very fast, the intermicellar exchange will be the controlling step. Thus a change in the proportion of reactants inside the micelles would weakly affect the reaction rate. In contrast, the repercussions of intermicellar exchange rate on slower reduction metal strongly depend on the reactant proportion due to a cage-like effect. In this way, the reaction rate of the slower metal can be manipulated by changing reactant proportions. As a nanoparticle is build up by bringing together new layers according to the order of metals reduction, the relative reaction rates of both metals determine the metals arrangement. As a result, maintaining the average concentration and microemulsion composition, the metal distribution in a Au/Pt bimetallic nanoparticle can be manipulated just by varying the precursors proportion. For example, a core–shell structure is obtained using a proportion 1:1, a thick Pt shell with an enriched Pt core is obtained using higher Pt proportion, and pure Au covered by a mixed Pt-Au shell results if the precursors are mostly Au salts. These conclusions can also be applicable to other metals whose difference between standard reduction potentials is about 0.2–0.3 V. These results are very promising for the design of the experimental conditions to obtain a nanoparticle with a specific distribution of metals.

ACKNOWLEDGEMENTS

We want to acknowledge the financial support of the MCI, Spain (MAT2011-28673-C02-02), and Xunta de Galicia (INCITE09314261PR).

REFERENCES

1. Boutonett, M.; Kizling, J.; Touroude, R.; Maire, G.; Stenius, P. *Appl. Catal.* **1986,** *20,* 163–177.

2. Sánchez-Dominguez, M.; Boutonnet, M.; Solans, C. *J. Nanopart. Res.* **2009,** *11,* 1823–1829.

3. Niemann, B.; Veit, P.; Sundmacher, K. *Langmuir* **2008,** *24,* 4320–4328.

4. Boutonnet, M.; Kizling, J.; Stenius, P.; Maire, G. *Colloids Surf., A* **1982,** *5,* 209–225.

5. Destrée, C.; Nagy, J. B. *Adv. Colloid Interface Sci.* **2006,** *123–126,* 353–367.

6. Capek, I. *Adv. Colloid Interface Sci.* **2004,** *110,* 49–74.

7. Yashima, M.; Falk, L. K. L.; Palmqvist, A. E. C.; Holmberg, K. *J. Colloid Interface Sci.* **2003,** *268,* 348–356.

8. Pileni, M. P. *J. Exp. Nanosci.* **2006,** *1,* 13–27.

9. Sánchez-Domínguez, M.; Aubery, C.; Solans, C. New Trends on the Synthesis of Inorganic Nanoparticles Using Microemulsions as Confined Reaction Media. In *Smart Nanoparticles Technology;* Hashim, A., Ed.; InTech: Rijeka, Croatia, 2012; pp 195–220.

10. Kuriyedath, S. R.; Kostova, B.; Kevrekidis, I. G.; Mountziaris, T. J. *Langmuir* **2010,** *26,* 11355–11362.

11. Touroude, R.; Girard, P.; Maire, G.; Kizling, J.; Boutonnet-Kizling, M.; Stenius, P. *Colloids Surf., A* **1992,** *67,* 9–19.

12. Boutonnet, M.; Lögdberg, S.; Svensson, E. E. *Curr. Opin. Colloid Interface Sci.* **2008,** *13,* 270–286.

13. Palmqvist, A. E. C. *Curr. Opin. Colloid Interface Sci.* **2003,** *8,* 145–155.

14. Chen, C.-H.; Hwang, B.-J.; Wang, G.-R.; Subramanya Sarma, L.; Tang, M.-T.; Liu, D.-G.; Lee, J.-F.*J. Phys. Chem. B* **2005,** *109,* 21566–21575.

15. Tao, F.; Grass, M. E.; Zhang, Y.; Butcher, D. R.; Aksoy, F.; Aloni, S.; Altoe, V.; Alayoglu, S.; Renzas, J. R.; Tsung, C.-K.; Zhu, Z.; Liu, Z.; Salmeron, M.; Somorjai, G. A. *J. Am. Chem. Soc.* **2010,** *132,* 8697–8703.

16. Yonezawa, T.; Toshima, N. *J. Chem. Soc., Faraday Trans.* **1995,** *91,* 4111–4119.

17. Deepak, F. L.; Casillas-García, G.; Esparza, R.; Barron, H.; Jose-Yacaman, M. *J. Cryst. Growth* **2011,** *325,*60–67.

18. Feng, J.; Zhang, C.-P. *J. Colloid Interface Sci.* **2006,** *293,* 414–420.

19. Wu, M.-L.; Chen, D.-C.; Huang, T.-C. *Langmuir* **2001,** *17,* 3877–3883.

20. Ferrando, R.; Jellinek, J.; Johnston, R. L. *Chem. Rev.* **2008,** *108,* 845–910.

21. Habrioux, A.; Vogel, W.; Guinel, M.; Guetaz, L.; Servat, K.; Kokoh, B.; Alonso-Vante, N.*Phys. Chem. Chem. Phys.* **2009,** *11,* 3573–3579.

22. Ohkubo, Y.; Shibata, M.; Kageyama, S.; Seino, S.; Nakagawa, T.; Kugai, J.; Nitani, H.; Yamamoto, T. A.*J. Mater. Sci.* **2013,** *48,* 2142–2150.

23. Angelescu, D. G.; Magno, L. M.; Stubenrauch, C. *J. Phys. Chem. C* **2010,** *114,* 22069–22078.

24. Lu, J.; Hong, L.; Yang, Y.; Xu, S.; Wang, S.; Yuan, J.; Niu, L. *J. Nanopart. Res.* **2013,** *16,* 2162–2171.

25. Wanjala, B. N.; Luo, J.; Loukrakpam, R.; Fang, B.; Mott, D.; Njoki, P. N.; Engelhard, M.; Naslund, H. R.; Wu, J. K.; Wang, L.; Malis, O.; Zhong, C.-J. *Chem. Mater.* **2010,** *22,* 4282–4294.

26. Hwang, B.-J.; Subramanyam Sarma, L.; Chen, J.-M.; Chen, C.-H.; Shih, S.-C.; Wang, G.-R.; Liu, D.-G.; Lee, J.-F.; Tang, M.-T. *J. Am. Chem. Soc.* **2005,** *127,* 11140–11145.

27. Silva, O. F.; Fernández, M. A.; Silber, J. J.; de Rossi, R. H.; Correa, N. M. *ChemPhysChem* **2012,** *13,* 124–130.

28. Baruah, B.; Roden, J. M.; Sedgwick, M.; Correa, N. M.; Crans, D. C.; Levinger, N. E. *J. Am. Chem. Soc.* **2006,***128,* 12758–12765.

29. Tojo, C.; de Dios, M.; López-Quintela, M. A. *J. Phys. Chem. C* **2009,** *113,* 19145–19154.

30. Cason, J. P.; Miller, M. E.; Thompson, J. B.; Roberts, C. B. *J. Phys. Chem. B* **2001,** *105,* 2297–2302.

31. Shukla, D.; Mehra, A. *Langmuir* **2006,** *22,* 9500–9506.

32. Natarajan, U.; Handique, K.; Mehra, A.; Bellare, J. R.; Khilar, K. C. *Langmuir* **1996,** *12,* 2670–2678.

33. Voigt, A.; Adityawarman, D.; Sundmacher, K. *Nanotechnology* **2005,** *16,* S429–S434.

34. Tojo, C.; Blanco, M. C.; López-Quintela, M. A. *Langmuir* **1997,** *13,* 4527–4534.

35. Tojo, C.; Blanco, M. C.; Rivadulla, F.; López-Quintela, M. A. *Langmuir* **1997,** *13,* 1970–1977.

36. Barroso, F.; Tojo, C. *J. Phys. Chem. C* **2013,** *117,* 17801–17813.

37. Buceta, D.; Tojo, C.; Vukmirovik, M.; Adzic, R. R.; Deepak, F. L.; López-Quintela, M. A. submitted.

38. Quintillán, S.; Tojo, C.; Blanco, M. C.; López-Quintela, M. A. *Langmuir* **2001,** *17,* 7251–7254.

39. Fletcher, P. D. I.; Howe, A. M.; Robinson, B. H. *J. Chem. Soc., Faraday Trans. 1* **1987,** *83,* 985–1006.

40. Kitchens, C. L.; Bossev, D. P.; Roberts, C. B. *J. Phys. Chem. B* **2006,** *110,* 20392–20400.

41. Housaindokht, M. R.; Pour, A. N. *Solid State Sci.* **2012,** *14,* 622–625.

42. Dunstan, T. S.; Fletcher, P. D. I. *Langmuir* **2011,** *27,* 3409–3415.

43. Davies, S. CHEN 610 - Applied Statistical Thermodynamics Course.http://inside.mines.edu/~dwu/classes/chen610/projects/Davies%20Summary%20Report.pdf (accessed Oct 15, 2014).

44. Lemyre, J.-L.; Lamarre, S.; Beaupré, A.; Ritcey, A. M. *Langmuir* **2011,** *27,* 11824–11834.

45. Niemann, N.; Rauscher, F.; Adityawarman, D.; Voigt, A.; Sundmacher, K. *Chem. Eng. Process.* **2006,** *45,* 917–935.

46. Ethayaraja, M.; Dutta, K.; Muthukumaran, D.; Bandyopadhyaya, R. *Langmuir* **2007,** *23,* 3418–3423.

47. Choudhury, S. D.; Kumbhakar, M.; Nath, S.; Sarkar, S. K.; Mukherjee, T.; Pal, H. *J. Phys. Chem. B* **2007,** *111,*8842–8853.

48. Turro, N. J.; Weed, G. C. *J. Am. Chem. Soc.* **1983,** *105,* 1861–1868.

Chapter 4

Role and adverse effects of nanomaterials in food technology

Lahir Yogendra Kumar

Department of Biophysics, University of Mumbai, Santacruz (E), Mumbai, India.

ABSTRACT

Food processing is related to the practice adopted by the food and beverages industries to change raw plant and animal materials in to 'ready to eat form'. Nanomaterials are applied to food technology with respect to their properties and predetermined set goals such as taste, flavor, shelf-life, appearance, likes and dislikes of the consumers. Nanomaterials have specific potentials depending on their physicochemical properties such as surface effect, small size effect, quantum size effect and quantum tunneling effect. These properties regulate their behavior in the biosystem for they may be either tolerated or be the cause of disturbance to biochemical and/or physiological homeostasis. Mostly nanomaterials have the ability to reach target tissues or organs where their counterparts fail to reach in the organism. Nanoparticles like zinc, calcium and silver are found to be biocompatible and antimicrobial in nature. Hence, these are used in the form of edible film incorporated with cinnamon or oregano oil in the packaging of food. Generally, polymers are incorporated with the nanomaterials and are used in food packaging and food processing. Food processing is aimed at good food quality and safe evaluation by improving food sensing and better nanostructured ingredients. These nanomaterials improve the flexibility and durability of the food contents. The nanomaterials enter the body along with the food products, beverages and other drinks for consumption. Understanding the mechanism involved in toxicity due to nanomaterials may provide necessary information about the nanomaterials which will act as guide lines for their appropriate use in food technology and its related aspects. This will also help to develop more innovative devises to meet the future challenges in food technology. This overview is intended to evaluate the extent of overall role of nanomaterials in food technology.

KEYWORDS

Nanomaterials, biocompatibility, food-packaging, food safety, food processing

1. INTRODUCTION

The food industry is one of the prime industries and a major target of application of nanomaterials. Food processing refers to the practice adopted by the food and beverages industries to convert the raw plant and animal materials in to ready to serve or eat form of food for consumers [1]. Media has played its significant role to build a strong public opinion in favor of the processed food products. As a result of this impact on public, the market has been infiltrated with huge number and varieties of processed food products [2]. Nanomaterials play an effective role in food preservation, packaging process and packaging material [3]. Candies, sweets, and chewing gums have gained popularity because of the use of nanomaterials [4]. This success is based on the sensitivity and efficiency of the products of nanoscience. Uses of nano-scale materials in the food industry are likely to improve the processed food [5]. Nanomaterials are being used as coated materials in treatment and as diagnostic tools [6].

Nanomaterials have unique properties such as surface effect, small size effect, quantum size effect and quantum tunneling effect. These have the ability to exist in aerosol, suspension, make them worthy of application in enhancing the utility of the food products. The surface and interfacial properties of nanomaterials can be modified with surfactants; these engineered nanomaterials exhibit flexibility. This facilitate in fabricating nanomaterials for various uses. Various types of nano-objects are involved in the formulation of nanomaterials such as nanoparticles, nanorods, nanotubes, nanocomposites, and nanofilm etc. These are used as per the need of applications in varied industries. Mostly nanopowder may include TiO_2, SiO_2, ZnO, CeO_2; colloidal dispersions and nano-suspensions involve gold, silver, silica etc. Food nano-sensing is related with food quality and safe evaluation while nanostructured ingredients help to improve food processing [7]. In a broadly sense, the nano-food processing is accomplished in three major aspects namely production (this pertains to over all conversion of raw food materials in to edible form), processing (it is related to modify the processed food with respect to taste, shelf life, protection against possible pathogens etc, as per the need of consumers) and to retain required nutritional values [8]. Agriculture and agrobased industries are the sources of raw materials for food industries. The involvement of nanomaterials in agriculture, food and food technology appears to be inevitable to meet the current demand of human population.

2. REVIEW

2.1 Applications of nanomaterials in food processing

Food cycle includes all phases of food industry which are exploited in bringing food from field to table. It includes agropractice, post harvest practice and processing of raw food in to ready to eat-form. Agro-practice aspect is out of the current discussion but can be envisaged that use of pesticides, fertilizers on crops etc., are likely to be facilitated by some of the nanodevices. Post harvest aspect includes storing, aggregating the agriculture produce while processing of raw food deals with procuring grains, vegetables, fruits, and other ingredients in to processed food or ready to eat-form. Food nanotechnology goes a few steps ahead in making food palatable and its pleasant aftereffects by using nanomaterials in different forms at various stages of food processing.

Broad outline of the various stages of processing of food are packaging, designing and formulation, keeping the food fresh and usable form during storage, transportation, supply and disposal of the unused products beyond expiry period or spoiled food. The role of nanotechnology in these phases of the food industry has been established; it is becoming prominent, demanding and enduring day by day. Food processing implies that raw food is transformed in to pleasing, palatable, digestive, nutritive and hygienic form that can be consumed by humans. Elementarily, food processing remains relatively same though food and feeding habits are different in different parts of the world. The food processing has undergone varied changes in due course of time in accordance to the needs of the individual, life style and society. Application of the nanomaterials in food and food technology is based on the behavior of the nanoparticles; this behavior is primarily dependent on the scientific evidences obtained during the primary research. Ag and ZnO NPs play an effective role in maintaining the shelf-life of fresh orange juice and have recommended these NPs as safe material for food technology [9]. The major applications of nanotechnology in food engineering are as food nanosensing and nanostructured ingredients. Nanosensing pertains to better food quality and safety evaluation while nanostructured ingredients help to improve food processing. Improved food processing includes appropriate delivery of nutrients, bio-separation of proteins, fast sampling of biological, chemical contaminants and nanoencapsulation of nutraceuticals, stabilization, delivery and coloration of the food system [7]. Nanomaterials play an effective role to attain better encapsulation and release efficiency of active components of food in comparison to the traditional encapsulating agents. Some of the important food supplements are to be retained during the food processing and it is accomplished by using nanomaterials. Nanosized powders are used to elevate the degree of absorption of nutrients, nanoencapsulation of nutraceuticals which are used to increase absorption, better stability and targeted delivery, cellulose nanocrystal composites as drug carrier, vitamin spray dispersing active molecules as nanodroplets for efficient absorption and nanocochelates as coiled nanoparticles, are employed to ensure nutrients delivery to the cells without disturbing the color and taste of the food. Other varied modifications of nanomaterials include development of nanoemulsions, liposomes, micelles, bipolar complexes and cubosomes are effective in protection of bioactive ingredients, controlled delivery system, food matrix integration and masking the flavors that are undesirable [10]. Such manipulated enzymes are administered to enhance the bioavailability of essential nutrients that include essential minerals and vitamins and hydrolyze antinutrients. This practice of adding formulated enzymes ensures appropriate enzyme-substrate support system because this brings about higher surface to volume ratio as compared to traditional support system.

ZnO NPs are used in many nanostructures and have the ability to be fabricated in a wide range of morphological varieties [11]. Gelatinization of starch plays key role in processing of food based on starch because it influences the nutritional benefits (if starch used gets contaminated it can be the source of adverse effect) [12]. During the process of filtering beer from mother solution and procuring cheese from milk, nanoseives or nanomembranes are most commonly used [12]. Nanoparticles like zinc, calcium and silver are found to be antimicrobial in nature; hence, these are used in the form of edible film incorporated with cinnamon or oregano oil in the packaging of food. Nanofibers are fabricated from the shell of lobster and organic corn which are antimicrobial and can be easily biodegraded, thus, becomes preferred material for food processing. Prime concern of using nanomaterial is that the nano-additives should not carry insoluble biopersistant matter in the circulation of the consumer. Quite often food is spoiled because of oxidation of the organic and/or added components. A fabricated electrochemical biosensor (a redox biosensor) is able to detect phenolic compounds in the food within range 37nM to 4.33 μM [13]. Nanoencapsulates, silver and TiO_2 are among those nanomaterials that have been applied to the maximum extent [14]. Nanocomposites in the form of novel food, food or feed additives, biocides, pesticides and food contact materials have bright future in the field of food processing.

During food processing nanotechnology assures many benefits such as safety, security, quality and shelf life of the food products [15]. This interaction between nanomaterials and processed food can cause unknown and unexpected hazard to the consumer. Nanomaterials are used as nanocapsules to enhance bioavailability of nutraceuticals, standard ingredients like cooking oil, to encapsulate flavor enhancers, to infuse plant based steroids to replace cholesterol of meat; nanoemulsions and nanoparticles are employed to increase availability and dispersion of nutrients. Nanotubes and nanoparticles are used as gelation and viscosifying agents while nanoparticles are involved in selective binding and removal of chemicals and pathogens from food under consideration. The potential application of nanomaterial science is being carried out with reference to nanoscaled biomaterial, packaging, sensors to increase shelf life, to detect development of undesirable factors [16]. Inorganic nanomaterials are used in healthy food products [17]. Transition metals and metals like Ag, Fe, TiO_2, alkaline earth metals such as calcium, magnesium, and non-metals like silicates and selenium are extensively used in food processing. When children, at health center were fed ultra processed food, children aged 2-10 years, elder children (5-10 years) were found to consume relatively more ultra processed food in comparison to younger ones [18]. Silver zeolite is used as coated ceramic form; silver produced ROS which are responsible for the antimicrobial action [19]. Nanoiron is a suitable health supplement because it breaks down the contaminants and destroys microbes. Modified magnetic nanoparticles

(Fe_3O_4-SiO_2-NH_2) has been found to be relatively better agents for eliminating microbes and virus from water [20]. Further, nanomicelle based formulation involving natural glycerin is a very suitable option to remove pesticides residue from fruits and vegetables up to 99.9%.

2.2 Nanomaterials used in food packaging

In general nanomaterials used in food packaging include nanofilms as barrier materials to prevent spoilage and oxygen absorption, electrochemical nanosensors to detect ethylene, antibodies linked with fluorescent nanoparticles to detect chemical and food borne pathogens. Common packaging for processed food, beverages and other related products in which nanomaterials are exploited sandwich bags; salad bowl, carry bags, bear bottles, aluminum foils, etc. The polymers incorporated in the nanomaterials improve the flexibility and durability of the food contents. The engineered nanomaterials (ENMs) used nullify the effects of temperature, humidity and act as barriers for the potential organisms to harm or spoil the food thereby enhancing the stability of food. The ENMs incorporated with plastic polymers have antimicrobial properties that are used to monitor the condition of packed food. The fabricated packaging materials with such specific qualities are termed as improved nanocomposite/active nanocomposite and smart packaging [12]. When such types of engineered nanomaterials are used for food packaging there are chances of migration of the nanomaterial towards food, drinks, or any edible matter packed therein, this migration is prevented to a greater extent. The bottles or the containers to be used for packing drinks, beverages, and other edible material contains nanoclay composite; these composites are embedded in the layers of polyethylene terephthalate (PET). There are no detectable movements of nanoclay from PET to the food contents or drinks packed. Those engineered nanomaterials that have smaller size than 1nm could migrate through packaging because polymer matrices materials exhibit relatively lower dynamic viscosity and no interaction between the two [21]. When nanocomposite specifically involving silver and polyolefines were used for packaging then there seemed to be some chances of migration of very small nanomaterials but not the bigger ones. The bigger nanomaterials get bound to the polymer matrices such as polystyrene and polyethylene terephthalate because of the relatively higher dynamic viscosity. Very low degree of migration of Ag NPs occurred from containers made of polypropylene nanosilver-composite. There was no transfer of non-nanocomposite material from the intelligent, smart and active packaging fabricated nanotechnologically like inks, reactive nanolayers and analyte at nano scale. These packaging materials provide non-toxicity, safety standards as per the legislations, reliability of products and after- use disposal issues are taken care [12]. Research and development both are aimed to maintain specific safety standards keeping the expectations of consumers and to maintain the environment. The label on the packages are improvised with radiofrequency identification display (RFID), this technique is able to track the food product during its transportation and distribution. Antimicrobial activity of ZnO powder coated PVC film with respect to food pathogens revealed that electrostatic interaction between bacterial cell surface and nanoparticles inhibited the growth of pathogens [22]. Nanomaterials like CNT, silicon nanowires, ZnO-nanorods etc, elevated the detection capability of biomolecules [23]. Those nanomaterials that have small dimensions and unique physical, chemical and optical properties ensure the safety of food and other contents of processed products along with the detection of spoilage in packed conditions. Hence, these are most suitable and ideal as bioprobes and biosensors [13]. Nanomaterial such as cubosomes (5-10 nm), liposome (20 nm), micelles (5-500 nm) have the ability to encapsulate biopolymeric nanoparticles like starch, chitosan, polylactic acid, gum Arabic, carrageenan, and alginate etc, and fabricated to be used in smart packaging material [23]. The new generations of such nanoencapsulation system are double layered liposomes, solid-lipid-nanoparticles, colloidosomes, nanolaminates and composite nanofibers, incorporated while plastic polymers are combined with nanoclay to fabricate stronger and cleaner, flame resistant material with a higher degree of reliability. These materials meet the requirements of the manufacturer, consumers of the food and food technology.

Bayer Polymers have fabricated a film involving silicate nanoparticles. It prevents the movements of oxygen and other gaseous material and transfer of heat, retaining the moisture contents of food packed [12]. In the recent past nanoparticles like silver, ZnO, MgO are being used in beer packaging to prevent contamination of food and beverages because of their antimicrobial nature. Polylactic acid when incorporated with small concentration of silica nanoparticle becomes biodegradable material, cost effective, available commercially, and acted asa suitable packaging material [24]. If the resultant polymer matrix were incorporated with nanofillers its physicochemical properties were enhanced due to small size and increased area, its efficiency was observed to be increased many folds. Nanowheels are the devices which are bimetallic and in the form of 2D nanostructure. These are good materials to maintain heat and catalytic reactions of the food contents in packaging because these exhibit well-defined tunable surface resonance [25]. Food and crops are most vulnerable to microbial and fungal infection at any stage of food technological cycle and there is a dire need to protection where nanomaterials can be very useful [26,27]. Studied on antimicrobial activity of ZnO NPs with respect to *E. coli* and *Pseudomonas aeruginosa* and antifungal activity against *Aspergillus niger*, revealed that *maximum* inhibition was found to be at 100 µl, with concentration of ZnO NPs in 100µl was 100µg and 400µl respectively; because of the specific antimicrobial nature of ZnO NPs against food borne pathogens hence, these are used in food packaging and food preservation [28]. Nanoscaled electrospun fibers and encapsulated active compounds very effective in increasing the shelf life of the packed food [16]. Eugenol (a phenylpropene, an alkyl chain substituted guaiacol compound clear to pale yellow in appearance), it is a stable and preferred and successful preservative. Eugenol has antibacterial and antifungal properties. This compound is advantageous as it has the ability to regulate the release of the active encapsulated contents and is advantageous for effective pay-loads [29]. Nanocellulose (microfibrilated cellulose)–another biomaterial, used for packaging of food is preferred and other applications because the fabrication is relatively much easier than production of CNT [16]. Nanotechnology is involved in developing nanomaterials which have a specific degree of permeability, nanomaterials with enhanced barrier characteristics and mechanical features, antimicrobial and anti fungal surfaces and higher degree of biodegradability [19,30]. These new generations of nanomaterials are expected to meet most of the requirements of the future.

2.3 Fate of nanomaterials used in food-distribution and elimination

The nanomaterials enter the body of consumer along with food, beverages and other drinks. The ingested nanomaterials along with food undergo physiological changes. Some of these are likely get absorbed either directly or with absorbed food components. Size, shape and structure predispose them to cross the biological barrier of the gastrointestinal tract. These nanomaterials remain in the system either in bounded or unbounded form at least for some duration. During this period these interact with biomolecules either beneficially or adversely. Their elimination from the body is brought about by different modes depending on their interaction with biomolecules and physicochemical features. Specifically the engineered nanomaterial designed for specific purpose may pose a problem during their elimination from the body depending on the type of fabrication and material used.

The investigation related to elimination of quantum dots *in vivo* indicated that the behavior of QDs depended on the size, and surface chemistry. Those QDs which had coating of cysteine and 5.5nm diameter were eliminated through kidney while QDs with different dimension and surface properties need further investigation [31-33]. Opsonization (a process that prepares foreign materials or marks the pathogen for phagocytosis) occurs depending on their size and surface features during the process of elimination of nanomaterials [34]. Orally administered nanomaterials get absorbed and move across gastrointestinal tract and distributed in the body parts like kidney, liver, spleen, brain and respective target organs. Most of the orally administered nanomaterials were eliminated along with the feces and/or urine [35]. Higher concentrations of nanomaterials were excreted via kidney, hepatobiliary pathway. When neural and neuronal cell lines were exposed to carbon nanotubes, these grew and were better differentiated [36,37]. The nanomaterials were found to be cleared by macrophages by phagocytosis; the phagocytosis was enhanced by chemotactic attraction at least in the alveolar zone [35,38]. Clearance of nanomaterials occurred as soon as these materials reach various organs of the body or circulatory system; clearance process was accomplished by live and phagocytotic process of macrophages in spleen and other organs [39]. When nanomaterials like polysorbate (used in food and pharmaceuticals) 80-coated PBCA and pegylated PLA immunonanoparticles, were administered intravenously, could move through blood brain barrier and were found to be accumulated in the brain tissue resulting in neurotoxicity due to the respective physicochemical features [40]. ZnO nanoparticles (50µg/g wt of egg, 50µg/g wt of egg) reach in the embryonic hepatic tissue within 24 to 48 hr when administered in chorio-allantoic fluid in the egg (chicken) on 18/19 day during the incubation [41]. The human red blood cells exhibited viability and hemolytic response against ZnO nanoparticles (20µg/ml, 30µg/ml and 40µg/ml) [42]. ZnO nanoparticles, 20 nm and 70 nm sized ZnO NPs were administered orally in male and female rats. The concentration in plasma, liver, lung, and kidneys was found to be increased within 24 hr after administration following dose-dependent pattern. Degree of renal elimination of these nanoparticles was very less as compared to fecal elimination. ZnO NPs were unevenly absorbed in to blood circulation form gastrointestinal tract [43]. The excretion of TiO_2 and ZnO nanoparticles in Spargue-Dawley rats using dose (1041.5 mg/Kg body weight, for 13 days). The tissues considered for excretion were blood, liver, kidney, spleen, brain, and urine and feces. TiO_2 was observed to be less in these tissues in comparison to ZnO nanoparticles. ZnO NPs were observed to be very low in brain and spleen as compared to ZnO NPs; TiO_2 NPs exhibited lower renal excretion was in comparison to ZnO NPs while ZnO NPs were excreted in greater amount via urine but in feces it was very low in comparison to TiO_2 NPs [44]. Intestinal goblet cell secretion path way is one of the modes to eliminate the nanoparticles from the biosystem. This was investigated by injecting 30-200 nm sized activated carbon nanoparticles in the yolk of Zebra fish, these nanoparticles were excreted directly in the lumen if intestine and hepatobiliary path was not preferred [45].

Most of the nanomaterials present in circulation in the body are cleared by reticuloendothelial cells. Those nanomaterials which degrade slowly and show a low degree of renal clearance get accumulated in liver and spleen. Nanomaterials having size around 5 nm and present in blood circulation undergo renal clearance. Those having 10-20 nm are separated by liver; 200nm sized NPs get picked up by Kupffer's cells and sinusoidal spleen. Others are taken care of by macrophages, opsonin assisted NMs undergo phagocytosis [46]. All of these varied observations related to various nanomaterials prove that most of these get distributed in biosystem and also follow a specific path for their elimination irrespective of the selected and intended purposes.

2.4 Evaluation of nanomaterials before theiruse in food technology

There have been tremendous applications of nanomaterials in the occupational, environmental and consumer sectors. This has elevated the public concern with respect to adverse effects and health hazards. This concern is expressed as risk factor in consumers. Intense use of engineered nanomaterials in almost every aspects of human life has raised the concern in the minds of the consumers. The nanomaterials at the work place are effective and likely to cause adverse effects (poor working conditions or because of the negligence) either immediately or at a later stage of life. Worker may suffer from the derogative effects of these nanomaterials (either acute or chronic toxicity) depending on his/her general health. The worker may experience this impact either during the tenure of the job or latter part of the life.

The toxicity of pristine and functionalized carbon nanotubes were observed from mild to moderate levels of toxicity; there are many issues that demand for research approach complementary to *in vivo* investigation related to understanding the biochemical, physiological aspects and health hazard related to nanomaterials [47]. There are specific techniques to engineer safe nanomaterials for the varied uses [48]. There is a need to understand the risk and beneficial aspects of the nanomaterial to be used. This whole approach should be nanorelevant. During the analysis one should consider the effect of nanomaterials with respect to heterogeneity, morphology, complexes formed (after nanomaterial is mixed with food). Thus, there should be clear protocol related to collection of the sample, strategies concerned with consumed food, type of processed food, amount of food consumed and the movement of food in the alimentary canal from where food sample is to be procured for analysis [49]. These precautions will ensure the safety from the nanomaterials while being exploited in the field of food technology and related agrobased industries. The evaluation of safe use of nanomaterials is of great significance to ensure most of the beneficial effects rather than adverse. Nanomaterials like ZnO, Ag, Au, TiO2, quantum dots, dendrimers, nano-prosthetic, nanocomposite etc, if judicially exploited will result in beneficial impact. Ampelopsin is a major flavonoid, a potential antioxidant that prevents rancidity in diets; hence, can be used in food products [50].

2.5 Tolerernce of nanomaterials used in food technology

Nanomaterials to be used in food technology should be tolerant to the consumer and if these result in adverse impact their use in food products is futile. Biocompatibility of nanoparticles is their ability to coexist within the body of a biosystem without disturbing the biochemical or molecular integrity of cell or tissue. Biocompatibility is regarded as the 'ability to be in contact with living system without causing adverse effects or biohazards are safely manipulated,' [51,52]. Compatibility has been redefined as the ability of a chemical to carry on its designed function with respect to a biochemical/medical therapy without causing any local or systemic adverse effect in the body of the recipient and have potential to create beneficial cellular and/or tissue response in the given conditions with optimum performance [53]. Biosystems include microbes, wild varieties and domestic organisms; domestic organisms encompass agriculture crops, poultry, animal husbandry, organisms of aesthetic value and others used for scientific research. This ability of coexistence depends on the interaction between the specific nanomaterial and the specific species; this interaction is likely to be varied depending on the specie specificity. Biocompatibility may show fluctuations, beneficial and derogative state; these associations are related to nurturing the survival of a biosystem in given environmental conditions.

The compatibility of nanomaterial depends on the dose, type, size, physicochemical nature, type of the tissues, number and the duration of exposure and the photoperiod. Regulated lower doses of nanomaterials are preferable for survival and to evaluate their biocompatibility in a biosystem. It is appropriate to avoid those interactions which disturb the biocompatibility such as surface oxidation, photosensitivity etc, to ensure appropriate tolerance. The biocompatibility can be ensured or enhanced by taking preventive steps such as coating the suspected surface, evaluating photo-effect (incase photosensitive nanomaterials) etc, of the nanomaterials under consideration. Zinc is one of the inorganic elements which are needed by human body; zinc oxide is non-toxic to man but it is toxic to microorganisms. ZnO nanoparticles are biocompatible to man, hence, can be suitable in the role of biosensors [54]. Biocompatibility of nanomaterials can be enhanced by self assembly with chitosan and modified lecithin [55]. Biocompatible gold nanoparticles can be formulated with green chemistry approach [56]. Inorganic nanoplate for of nanomaterials can be used as theranostics for cancer [46].

2.6 Biological adverse effects due to nanomaterials used in food

Nanomaterials are used in abundance in food technology. The processed food products have been infused with nanomaterials. These carry nanomaterials in the gastrointestinal tract of the consumers. These in turn undergo physiological changes. As a result there is every possibility that these nanomaterials are likely to interact with the tissue, cellular organelles, and biomolecules. This interaction may be beneficial and/or adverse physiologically, biochemically, cytologically, histologically or genetically. Some of the nanomaterials like metallic and metallic-oxides nanoparticles, quantum dots, fullerenes, and fibrous nanomaterials have been found to cause adverse effects that include fragmentation of chromosome, breakage of strands of DNA, changes in gene precision etc. There are reports on clinical toxicity but major studies indicate that nanomaterials are able to initiate adverse biological interactions leading to untimely physiological disruption and toxicological out come in due course of time.

Silica nanoparticles cause oxidative stress and resulted in cytotoxicity *in vitro* and *in vivo*; elevated lipid peroxidation, LPO, ROS, and declined level of cellular glutathione [57,58]. There is every possibility of known and unknown varied hazards based on systemic investigation on cells treated with nanomaterials with respect cytotoxicity and inflammation [59]. AuNPs and Ag NPs have potential to caused depolarization of α-tubulin-major component of microtubule, affecting adversely the cellular structure and associated cytoskeleton of the cell [60]. The mechanism involved in the interaction between nanomaterials and living biosystems is still ambiguous [61]. The efficacy of nanomaterials with which these interact with biological molecules in the ambient atmosphere and the formation of varied end products is highly complex phenomenon. This complexity becomes more challenging with the slight changes brought about in the nanomaterials due to coating, charge, size, surface properties, dispersion, agglomeration, concentration and matrix.

Nanomaterials have the potential to cause mutagenicity, clastogenicity, aneugenicity (causing the possession of abnormal chromosome numbers), apoptosis, autophagy and mitoptosis (death of mitochondria) etc, [6]. There is lot of toxicological data related to cytotoxicity, genotoxicity, repeated dose toxicity and biokinetics are available related to silica, titanium dioxide, and silver; there are limited records concerned with the use of nanomaterials in agriculture, food, feed application [14]. The resume of legislation and regulations in European and non-European countries have adopted EU legal acts and formulated the guide lines to be followed for the related industries.

Mostly iron oxide nanoparticles enter mammalian cells via endocytosis, sometimes by pinocytosis, through reticuloendothelial system, while in blood circulation opsonization of nanoparticles to cell surface was also involved. Internalization of iron oxide nanoparticles was observed to be directly related to their size [62]. Graphene induced toxicity and apoptosis in cell culture has been observed; this adverse response was found to be in relation to dose and duration dependent. Graphene (100-110nm) also caused cytotoxic state and damage to mitochondria in case of neural cells also [63]. Carbon nanotubes cause varied toxic response depending on the type and the applications [64]. The toxicity of dendrimers depended on their size and shape as these nanomaterials exhibit closeness/resemblance to the components of the cells like DNA, proteins etc. Their structural aspects were found to be very suitable to provide space for encapsulation of food ingredients and therapeutic drugs, imaging molecules and conveniently regulated the release. This specific feature was also able to reduce toxicity. However, these nanomaterials exhibit hemolytic and hematological toxic responses in a given biological system. Their features such as cationic surface charge and the negative charge on the biological membrane interact leading to the toxicity *in vivo* [65-67]. To counteract this adverse effect these are engineered to be biocompatible and biodegradable. During the experimentation with quantum dots, the observation indicated oxidative stress; influencing signaling pathway between macrophages. The interaction of QDs was found to be dose dependent and exhibited adverse effects on cell division and the growth of the cell [68]. Engineered nanoparticles relevant to various industries like SiO_2, ZnO, Fe_2O_3, Ag, CeO_2, had exhibited dose dependent elevated cytotoxicity and damage of DNA, affected cell proliferation in human lymphoblastoid and adherent ovary cell culture [69].

The metal oxide NPs, metallic NPs, quantum dots, SiO_2 NPs, CNT and other nanomaterials with respect to size, shape, surface charge, type of coating material, reactivity, dose, exposure, route of administration exhibit toxic behavior. Further, it was found that stable metallic oxides, mostly have not shown toxicity but metallic NPs bring a change in reduction potential which causes cytotoxicity and genotoxicity. Those NPs which were in soluble form were definitely found to be toxic in comparison to non-soluble [70]. C60 got linked with the DNA at its minor grooves and initiated unwinding; when it got linked with RNA it inhibited replication of DNA [71]. When C60 bound to nucleic acid the range of energy involved varies within -56 to -10 $kcl.mol^{-1}$ depending on the different genomic fragment [71]. There are views with respect to optimization of immobilized enzymes, immunodiffusional limitations, maximum surface area per unit mass and high effective enzyme loading. Further, nanoparticles affect the behavior of the enzymes in transitional region/phase between heterogeneous (immobilized) enzymes and homogeneous (with soluble free enzyme) enzyme catalysis. These nanoparticles influenced the reaction particle mobility related to the particle size and solution viscosity [72]. Polystyrene particle when used on an enzyme model α-chymotrypsin it got attached to NPs covalently up to 6.6 wt%. Factors such as change in particle size, viscosity of the reaction medium affect the mobility of the enzyme catalyst; there is a change in the intrinsic activity of the said enzyme (the particle attached enzyme), Km, Kcat were found to be affected because of the particle size and viscosity [72]. Structure and function of an enzyme can be changed by nanoparticles;

the enzyme–nanomaterial interaction is governed by the key properties such as structure, size, shape, surface chemistry and charge. This is because of the electrostatic complementarities between carboxylate end group and cationic residues present around the periphery of the active site [73].

When the surface of the nanoparticles was modified, these NPs affected the enzyme function in comparison to traditional materials. This factor could be modified with organic molecules via covalent or non-covalent modification to prepare functional nanoparticles. Suitably functionalized nanoparticles posses some specific abilities like preventing non-specific binding and recognizing specific biomacromolecules; this aspect of modified nanomaterials could specifically regulate enzyme activities [48]. The effects of iron oxide NPs on tissue and the enzymes of live in rats, caused hepatotrophic effects; the levels of alkaline phosphatase, aspartate amino transferase and alanine amino transferase enzyme activities were elevated [73]. Single enzyme NP of celluloclast B G, cellulose enzyme (SEN-CK) has the ability to degrade great substrate like cellulose and the stability of the SEN-CK enzyme has been found to be significantly better than that native enzyme [74].

Moderate doses of gold, silver and ZnO NPs (50 ppm, average size of NP10nm; 25, 50 100ppm) elevated LDH activity with all concentrations of the three nanoparticles used [75]. Specific and unique features of nanomaterials are the factors that affected toxicity and their interference during the assessment processes. Nature of nanomaterials and the mode of the behavior of the respective nanomaterials in the biosystems remain ambiguous in the current scenario of the toxicity assessment pattern [76]. One of the physiological and biochemical disturbances caused is related to change in the redox potential in the affected cell/tissue; reactive species of oxygen and nitrogen caused lipid oxidation and this condition can be created by nanomaterials [77]. SPION showed uniform distribution endothelial cells; this distribution was affected by the parameters like shape, size and other geometrical aspects and conditions related to hemodynamics [78]. Nature, type of coating on quantum dots the major parameters that played an effective role in their toxic activity; features like size, charge and the surface coating are the deciding factor of their uptake, mechanism of internalization and toxic behavior Table 1 [79].

Table 1 : Some of the adverse responses of nanomaterials.

Nanomaterials and parameters	Tissues/cells and Responses	References
CNT (single walled C-nanotube; 24 mg/ml	Plasma membrane destabilized, declined viability, oxidative stress induced and caused change in ultra structural morphological changes	[80]
Single walled C-nanotubes 200μg/ml	Inhibited cell proliferation, induced apoptosis, reduced the ability of cell adhesion, viability loss of embryonic renal cells	[81]
Single walled C-nanotubes and multiwalled nanotubes, 22.6μg/ml, 226μg/ml	High toxicity in macrophages in alveolar tissue, loss of viability, detachment of cells and change in protein expression	[82]
Single walled nanotubes 24 hr	Morphological changes in lamellar structures, microvilli, tonofilaments and desmosome junctions in relation to cell membrane	[83]
Al_2O_3, CeO_2, TiO_2 metal oxides nanoparticles	Change in membrane potential, induce lipid peroxidation and oxidative stress , cytotoxicity was observed,	[84]
Cationic nanoscale materials (related to surface charge)	Spherical NPs were taken up 500% more readily than rod shaped NPs, internalized NPs bombarded the endosomes and reached cytosol machinery, cationic NPs move across the cell membrane by forming transient opening and due to hydrophobicity, induced cytotoxicity	[85]
Silica NPs	Resulted cytotoxicity, caused oxidative stress in vivo and in vitro, elevated lipid peroxidation resulting rise in ROS and decline in cellular glutathione (GHS)	[58,86]
Gold NPs (AuNPs), Silver NPs (AgNPs), CoCrNPs	Depolarization of α-tubulin, disturbing the cellular structure, partial fragmentation of Protein	[60]
Silver zeolite NMs	Produces reactive oxygen species	[19]
Carbon nanotubes	Causes inflammation in neural cells (glial cells), cytokine	[87]
Dextran coated SPION(20nmand 60 nm)	Induced proinflammatory cytokine	[86]
Colloidal gold NPs (1μM/5μL)	Affected electrical activity in retina but no morphological changes	[87]

3. CONCLUSION

Nanotechnology is a multi-interdisciplinary technology having multifaceted applications. Its products are exploited in the formulation of personal care, health care, commercial and industrial applications. This ability is in relation to the structure, function and functionalization of either some groups of a molecule or molecule as a whole. Nanomaterials are being exploited in the processing of raw food to 'readyto-eat state'. This journey of food involves conversion of raw food in to prepared food (preprocessing), processing, packaging, transporting, supply, storage and disposal of expired or spoiled food and the packaging material. Nanomaterials are applied to food technology with respect to their properties and predetermined set goals such as taste, flavor, shelf-life, appearance, likes and dislikes of the consumers and the financial and commercial aspects. Prior to use of the nanomaterial it is quite essential to evaluate their biocompatibility to establish their utility in food or the consumer's intention to attain and maintain good active health. Processing and packaging is an essential aspect; its efficiency and effectiveness depends on the material used to pack the food to keep it fresh, usable, retention of nutritional values for the varied age groups. Biosensors are the devices which help to detect any derogative changes, infection or degree of spoilage of food packed. It is quite important to know distribution, fate and elimination of the nanomaterial from the body of the consumer. Various forms of nanomaterials devised, fabricated such as nanoparticles, nanofibers, nanomembranes, nanotubes, liposomes, nanofilms, etc, are used in food and food technology. It is very essential to estimate the over or abuse of these materials to avoid adverse effects. Study related to nanotoxicity is an appropriate approach to find an optimum, beneficial doses of nanomaterials to be used and their elimination from the biosystem. Even though it is claimed that the amount used of these products is within safe limits but the amount released directly or indirectly in the environment is comparatively much higher. Perspective of use of nanomaterials lies in the hands of personals involved in the field of nanotechnology, food technology and those trying to understand the adverse effects. The insight of the mechanism involved in interaction of nanomaterials and biosystems will guide the appropriate and effective manipulation for the good of man-kind and environment.

COMPETING INTERESTS

The author declares that he has no competing interests.

REFERENCES

1. Monterio C R and Levy R B. **A new classification of food based on extent and purpose of their Processing.** *Public Health Nutrition*. 2011; **25**:2039-2049.

2. Kathleen S, Christopher M, Lynn L B, Marilyn H, Chen J, Henry K and Jane H. **Nanotechnology and the environment.** C R S Press, Taylor and Francis Group, Boca Raton, London, New York. 2009.

3. Sastry R R, Rashmi H B and Rao N H. **Nanotechnology for enhancing food security in India**. *Food Policy.*2011; **36**:391-400.

4. Alex W, Paul W, Lars F, Kiril H and Von Goetz N. **Titanium di oxide in food and personal Care products.***Environ Sci Technol*. 2012; **6**:2242-2250.

5. Sanguansri P and Augustine M A**. Nanoscale materials development: food industry perspective**. *Trends Food Sci Technol*. 2006; **17**:547-556.

6. Jennifer M and Wnuk M. **Nanoparticles technology as a double edged sword: cytotoxic: Genotoxic and epigenetic effects on living cells.** *Nanobiotech.* 2013; **4**:53-63.

7. Ravichandran R. **Nanotechnology application in food and food processing: innovative green approaches, opportunities and uncertainties for global market**. *Int J Green Nanotechnology, Physics and Chemistry.*2010; **1**:P72-P96.

8. Timothy V D. **Application of nanotechnology in food packaging and food supply: Barrier materials, antimicrobial and sensors**. *J Colloid Interface Sci.* 2011.

9. Emamifar A, Kadivar M, Shahedi M and Zad S S. **Evaluation of nanocomposite packaging containing Ag and ZnO NPs on shelf-life of**

fresh orange juice. *Innovative Food Sci and Emerging Technol.* 2010;**11**:742-748.

10. Chaudhry Q and Castel L. **Food application of nanotechnologies: an overview of Opportunities and challenges for developing countries**. *Trends Food Sci Technol*. 2011; **22**:595-603.

11. Huang X, Li Y, Chen Y, Zhou H, Duan X and Huang Y. **Plasmonic and catalytic AuPd nanowheels for the efficient conversion of light into chemical energy**. *Angew Chem Int Ed Engl*. 2013; **52**:6063-7.

12. Sharon M, Sharon M, Pandey S and Oza G. **Biotechnology concept and application**. Ane Books (P) Ltd, New Delhi, Mumbai. 2012.

13. Rather JA, Pilehvar S and De Wael K. **A biosensor fabricated by incorporation of a redox mediator into a carbon nanotube/nafion composite for tyrosinase immobilization: detection of matairesinol, an endocrine disruptor**. *Analyst*. 2013; **138**:204-10.

14. RIKILT and JRC. **Inventory of nanotechnology application in agriculture, feed and food Sector**. *FSA supporting publication.* 2014; 125.

15. Sahu J K. **Introduction to advances food process engineering**. CRS Press, Taylor and Francis Group, Boca Raton, Fl, USA. 2014.

16. Chen H, Seiber JN and Hotze M. **ACS Select on nanotechnology in food and agriculture: a perspective on implications and applications**. *J Agric Food Chem*. 2014; **62**:1209-12.

17. Malik A, Erginkaya Z, Ahmed S and Erten H. **Food processing: strategies for quality assessment**. Springer, New York, London. 2012.

18. Sparrenberger K, Friedrich RR, Schiffner MD, Schuch I and Wagner MB. **Ultra-processed food consumption in children from a Basic Health Unit**. *J Pediatr (Rio J)*. 2015.

19. Berekaa M M. **Nanotechnology in food industry: advances in food processing, packaging and food safety**. *Int J Curr Microbiol Appl Sci*. 2015; **4**:345-357.

20. El-Mashad H M and Pan Z. **Food decontamination using nanomaterials**. *MOJ Food Processing and Technology*. 2015; **2**:00011.

21. Simon P, Chaudhry Q and Bakos D. **Migration of engineered nanoparticles from polymer packaging of food–physicochemical review**. *J Food Nutri Res.* 2008; **47**:105-113.

22. Li X, Xing Y, Jiang Y, Ding Y and Li W. **Antimicrobial activities of ZnO powder coated PVC film to inactivate food pathogens**. *Int J Food Sci Technol.* 2009; **44**:2164-2168.

23. John D F. **New and emerging applications of nanotechnology in food supply**. 2010**.**

24. Tanja I R, Banka M P, Ivan S R, Oskar J B, Sebastian B and Novakovic A R. **The influence of silica NPs on mechanical properties of polylactic acid as a food packaging material**. *Food and Feed Research.* 2012;**39**:97-102.

25. Huang X, Li Y, Chen Y, Zhou H, Duan X and Huang Y. **Plasmonic and catalytic AuPd nanowheels for the efficient conversion of light into chemical energy**. *Angew Chem Int Ed Engl.* 2013; **52**:6063-7.

26. Lahir Y K. **Toxicology**. Agrobios (India), Jodhpur, India. 2009; 55-94.

27. Lahir Y K. **Principles and applications of Toxicology**. *See Kay Publications, Bangalore, India*. 2013; **45**:171-172.

28. Chitra K and Annadurai G. **Antimicrobial activity of wet chemically engineered spherical shaped ZnO NPs on food born pathogen**. *Int Food Res J.* 2013; **20**:59-64. | Pdf

29. Kamatou GP, Vermaak I and Viljoen AM. **Eugenol--from the remote Maluku Islands to the international market place: a review of a remarkable and versatile molecule**. *Molecules*. 2012; **17**:6953-81.

30. Othman S H. **Bionano-composite materials for food packaging applications**. *Agri Sci Proceedings*. 2014;**2**:296-303.

31. Wang H, Wang J, Deng X, Sun H, Shi Z, Gu Z, Liu Y and Zhao Y. **Biodistribution of carbon single-wall carbon nanotubes in mice**. *J Nanosci Nanotechnol*. 2004; **4**:1019-24.

32. Smart S, Cassidy A T, Lu G Q, and Martin D J. **Biocompatibility of CNTs**. *Carbon.* 2006; **44**:1034-1047.

33. Fischer HC and Chan WC. **Nanotoxicity: the growing need for in vivo study**. *Curr Opin Biotechnol*. 2007;**18**:565-71.

34. Curtis J, Greenberg M, Kester J, Phillips S and Krieger G. **Nanotechnology and nanotoxicology: a primer for clinicians**. *Toxicol Rev*. 2006; **25**:245-60.

35. Hagens WI, Oomen AG, de Jong WH, Cassee FR and Sips AJ. **What do we (need to) know about the kinetic properties of nanoparticles in the body?** *Regul Toxicol Pharmacol*. 2007; **49**:217-29.

36. Galvan-Garcia P, Keefer EW, Yang F, Zhang M, Fang S, Zakhidov AA, Baughman RH and Romero MI. **Robust cell migration and neuronal growth on pristine carbon nanotube sheets and yarns**. *J Biomater Sci Polym Ed*. 2007; **18**:1245-61.

37. Jan E and Kotov NA. **Successful differentiation of mouse neural stem cells on layer-by-layer assembled single-walled carbon nanotube composite**. *Nano Lett*. 2007; **7**:1123-8.

38. Oberdorster G, Maynard A, Donaldson K, Castranova V, Fitzpatrick J, Ausman K, Carter J, Karn B, Kreyling W, Lai D, Olin S, Monteiro-Riviere N, Warheit D and Yang H. **Principles for characterizing the potential human health effects from exposure to nanomaterials: elements of a screening strategy**. *Part Fibre Toxicol*. 2005; **2**:8.

39. Forbe T, Garcia M, and Gonzalez E. **Potential risk of NPs**. *Cienc Technol Aliment*. 2011; **31**. | Pdf

40. Clift MJ, Varet J, Hankin SM, Brownlee B, Davidson AM, Brandenberger C, Rothen-Rutishauser B, Brown DM and Stone V. **Quantum dot cytotoxicity in vitro: an investigation into the cytotoxic effects of a series of different surface chemistries and their core/shell materials**. *Nanotoxicology*. 2011; **5**:664-74.

41. Pardeshi P, Nawale A B, Mathe V L, Lahir Y K and Dongre P M. **Effects of zinc oxide nanoparticleson the hepatic tissue of chicken embryo: A histological approach**. *Bionano Frontier*. 2014; **2**:176-180. | Pdf

42. Shirsekar P P, Kanhe N, Mathe V L, Lahir Y K and Dongre P M. **Interaction of zinc oxide Nanoparticles with human red blood cells**. *Bionano frontier*. 2015.

43. Baek M, Chung HE, Yu J, Lee JA, Kim TH, Oh JM, Lee WJ, Paek SM, Lee JK, Jeong J, Choy JH and Choi SJ.**Pharmacokinetics, tissue distribution, and excretion of zinc oxide nanoparticles**. *Int J Nanomedicine*. 2012; **7**:3081-97.

44. Cho W-S, Kang B-C, Lee J K, Jeong J, Che J-H and Seok S-H. **Comparative adsorption, distribution and excretion of TiO2 and ZnO NPs after oral administration**. *Particle Fiber Toxicology*. 2013; **10**:9.

45. Zhao B, Sun L, Zhang W, Wang Y, Zhu J, Zhu X, Yang L, Li C, Zhang Z and Zhang Y. **Secretion of intestinal goblet cells: a novel excretion pathway of nanoparticles**. *Nanomedicine*. 2014; **10**:839-49.

46. Huang X C, Luo Y L and Hsu H S. **Redox-triggered biocompatible inorganic nanoplate Form for cancer theranostics, In; Biological and Pharmaceutical applications of nanomaterials. Ed.** Prokopovich, CRS Press, Taylor Francis group, Boca. 2015.

47. Coccini T, Manzo L and Roda E. **Safety evaluation of engineered nanomaterials for health risk assessment: an experimental tiered testing approach using pristine and functionalized carbon nanotubes**. *ISRN Toxicol*. 2013; **2013**:825427.

48. Wu W, Ichihara G, Suzuki Y, Izuoka K, Oikawa-Tada S, Chang J, Sakai K, Miyazawa K, Porter D, Castranova V, Kawaguchi M and Ichihara S. **Dispersion method for safety research on manufactured nanomaterials**. *Ind Health*. 2014; **52**:54-65.

49. Szakal C, Roberts SM, Westerhoff P, Bartholomaeus A, Buck N, Illuminato I, Canady R and Rogers M.**Measurement of nanomaterials in foods: integrative consideration of challenges and future prospects**.*ACS Nano*. 2014; **8**:3128-35.

50. Hou X, Zhang J, Ahmad H, Zhang H, Xu Z and Wang T. **Evaluation of antioxidant activities of ampelopsin and its protective effect in lipopolysaccharide-induced oxidative stress piglets**. *PLoS One*. 2014;**9**:e108314.

51. Vert M, Doi Y, Hellwich K H, Hess M, Hedge P, Kubisa P, Renado M and Schue F. **Terminology for biorelated polymers and applications, (IUPAC Recommendations**. *Pure App Chem*. 2012; **84**:377-410.

52. CDC- Medicine.

53. Williams D F. **On the mechanism of biocompatibility**. *Biomaterials.* 2008; **29**:2941-2953.

54. He L, Liu Y, Mustapha A and Lin M. **Antifungal activity of zinc oxide nanoparticles against Botrytis cinerea and Penicillium expansum**. *Microbiol Res*. 2011; **166**:207-15.

55. Chuah AM, Kuroiwa T, Ichikawa S, Kobayashi I and Nakajima M. **Formation of biocompatible nanoparticles via the self-assembly of chitosan and modified lecithin**. *J Food Sci*. 2009; **74**:N1-8.

56. Gurunathan S, Han J, Park J H and Kim J H. **A green chemistry approach for synthesizing biocompatible gold NPs**. *Nanoscale Res Letters*. 2014; **9**:246. |Wang F, Gao F, Lan M, Yuan H, Huang Y and Liu J. **Oxidative stress contributes to silica nanoparticle-induced cytotoxicity in human embryonic kidney cells**. *Toxicol In Vitro*. 2009; **23**:808-15. |

57. Christen V and Fent K. **Silica nanoparticles and silver-doped silica NPs induce Endoplasmic reticulum stress response and alter cytochrome P450 1A activity**. *Chemosphere*. 2012; **87**:423-434

58. DeVoll J R. **Nanoparticles toxicity**. *Aviat Space Environ Med.* 2010; **81**:152-163.

59. Raghunathan V K, Hawkins S, Case C P, Mann S, Davis S and Lane J. **Factors that may Affect the nanotoxicity of heard materials for surgical application, Characterizing potential, risk posed by engineered NPs**. *A second U K Govt. research report Programme project.* 2011.

60. Elsaesser A and Howard CV. **Toxicology of nanoparticles**. *Adv Drug Deliv Rev*. 2012; **64**:129-37.

61. Cortajareno A L. Otega D, Ocampo S M, Gonzalez-Gracia A, Couleaud P. Cristobal B-I and Angel A-S.**Engineering iron oxide nanoparticles for clinical settings**. *Nanomedicine*. 2014; **1**:2.| Pdf

62. Seabra AB, Paula AJ, de Lima R, Alves OL and Duran N. **Nanotoxicity of graphene and graphene oxide**.*Chem Res Toxicol*. 2014; **27**:159-68.

63. GonzaloR-G, Zimmermann B and Weil M W. **Nanotoxicity and life cycle assessment**. 1st attempt towards the determination of characterization factors for carbon nanotubes, 2nd Int Conference on structural

nanocomposite (NANOSTRUCTURES 2014), IOP Publishing, *IOP Conf. Series, Materials Sci and Engineering.* 2014; **64**:012029.

64. Austruc D and Chardac F. **Dendritic catalytic and dendrimers in catalysis**. *Chem Rev.* 2001; **101**:2991-3024. | Pdf

65. Bardi G, Tognini P, Ciofani G, Raffa V, Costa M and Pizzorusso T. **Pluronic-coated carbon nanotubes do not induce degeneration of cortical neurons in vivo and in vitro**. *Nanomedicine*. 2009; **5**:96-104.

66. Sahu S C and Casciano D A. **Hand book of nanotoxicology, Nanomedicine and stem cell use in Toxicology**. John Wiley and sons. 2014.

67. Zeng Q, Shao D, Ji W, Li J, Li C and Song J. **The nanotoxicity investigation of optical nanoparticles to cultured cells in vitro**. *Toxicology Reports.* 2014; **1**:137-144.

68. Watson C, Ge J, Cohen J, Pyrgiotakis G, Engelward BP and Demokritou P. **High-throughput screening platform for engineered nanoparticle-mediated genotoxicity using CometChip technology**. *ACS Nano*. 2014; **8**:2118-33.

69. Kumar V, Kumari A, Guleria P and Yadav SK. **Evaluating the toxicity of selected types of nanochemicals**.*Rev Environ Contam Toxicol*. 2012; **215**:39-121.

70. Xu X, Wang X, Li Y, Wang Y and Yang L. **A large-scale association study for nanoparticle C60 uncovers mechanisms of nanotoxicity disrupting the native conformations of DNA/RNA**. *Nucleic Acids Res*. 2012;**40**:7622-32.

71. Jia H, Zhu G and Wang P. **Catalytic behaviors of enzymes attached to nanoparticles: the effect of particle mobility**. *Biotechnol Bioeng*. 2003; **84**:406-14.

72. Badadi V Y, Najafi L, Najafi A, Gholami H, Zarji M E B, Golzadeh J, Amraie E and Ali S. **Evaluation of iron oxide NPs effects on tissue and enzymes of liver in rats**. *J Pharmac Biomed Sci*. 2012; **23**:1-4.

73. Nagy H E. **Comparison of the structure and stability of single enzyme NPs**. *Hungarian J Indust Chemist Veszprem*. 2009; **37**:123-130.

74. Negahdary M and Ajdary M. **The toxicity of gold, silver and ZnO NPs on enzyme LDH in male mice**.*Annual Res and Review in Biology*. 2014; **4**:1346-1352. |

75. Ong KJ, MacCormack TJ, Clark RJ, Ede JD, Ortega VA, Felix LC, Dang MK, Ma G, Fenniri H, Veinot JG and Goss GG. **Widespread nanoparticle-assay interference: implications for nanotoxicity testing**. *PLoS One*. 2014; **9**:e90650.

76. Lahir Y K. **Lipid oxidation in biological systems: Biochemical, Biological and Biophysical aspects**. *Global J B B*. 2015; **4**:224-233.

77. Matuszak J, Zaloga J, Friedrich R P, Layer S, Nowak J , Odenbach S, Alexiou C and Cicha I. **Endothelial biocompatibility and accumulation of SPION under flow conditions**. *J Magnetism Magnetic materials*.2015; **380**:20-26.

78. Soenen S J, Parak W J, Rejman J and Manshian B. **Intracellular stability of inorganic NPs; effects on cytotoxicity, particle functionality and biomedical applications**. *Chemical Rev*. 2015; **115**:2109-21135.

79. Shvedova AA, Castranova V, Kisin ER, Schwegler-Berry D, Murray AR, Gandelsman VZ, Maynard A and Baron P. **Exposure to carbon nanotube material: assessment of nanotube cytotoxicity using human keratinocyte cells**. *J Toxicol Environ Health A*. 2003; **66**:1909-26.

80. Cui D, Tian F, Ozkan CS, Wang M and Gao H. **Effect of single wall carbon nanotubes on human HEK293 cells**. *Toxicol Lett*. 2005; **155**:73-85.

81. Jia G, Wang H, Lei Y, Pei R, Yan T, Zhao Y and Guo X. **Cytotoxicity of carbon nanomaterials:single walled nanotubes, multi walled nanotubes**. *Environ Sci Technol*. 2005; **39**:1378-1383.

82. Davoren M, Herzog E, Casey A, Cottineau B, Chambers G, Byrne HJ and Lyng FM. **In vitro toxicity evaluation of single walled carbon nanotubes on human A549 lung cells**. *Toxicol In Vitro*. 2007; **21**:438-48.

83. Lin W, Isaac S, Huang Y, Znou X and Yinfa Ma. **Cytotoxicity and cell membrane depolarization induced by aluminium oxide NPs in human lung epithelial cells A549**. *Toxicol Environ Chem*. 2008; **90**:983-996.

84. Verma A, Uzun O, Hu Y, Han HS, Watson N, Chen S, Irvine DJ and Stellacci F. **Surface-structure-regulated cell-membrane penetration by monolayer-protected nanoparticles**. *Nat Mater*. 2008; **7**:588-95.

85. Wang F, Gao F, Lan M, Yuan H, Huang Y and Liu J. **Oxidative stress contributes to silica nanoparticle-induced cytotoxicity in human embryonic kidney cells**. *Toxicol In Vitro*. 2009; **23**:808-15.

86. Li X, Xing Y, Jiang Y, Ding Y and Li W. **Antimicrobial activities of ZnO powder coated PVC film to inactivate food pathogens**. *Int J Food Sci Technol.* 2009; **44**:2164-2168.

87. Bardi G, Nunes A, Gherardini L, Bates K, Al-Jamal KT, Gaillard C, Prato M, Bianco A, Pizzorusso T and Kostarelos K. **Functionalized carbon nanotubes in the brain: cellular internalization and neuroinflammatory responses**. *PLoS One*. 2013; **8**:e80964.

88. Olson J L, Raul V M, Nicole N, David A A, Mandava N and Stodlt C R. **Intraocular biocompatibility of AuNPs**. *J Nanomater Mol Nanotechnol.* 2013; **2**:2.

89. Ingrid L B and Frank A W. **NPs toxicity by gastrointestinal route: evidence and Knowledge gaps**. *Int J Biomed Nanosci Nanotechnol.* 2013; **3**:1-2.

90. Lam CW, James JT, McCluskey R, Arepalli S and Hunter RL. **A review of carbon nanotube toxicity and assessment of potential occupational and environmental health risks**. *Crit Rev Toxicol*. 2006; **36**:189-217.

91. Benedicte T, Ramune R, Haywest B, Parrisa S and Schiestl R H. **Titanium-di-oxide nanoparticles induce DNA damage and genetic instability in vivo in mice**. *Cancer Res.*2009; **69**:8784.

92. Wu Z, Bin Z and Yan B. **Regulation of enzyme activity through interaction with NPs**. *Int J Mol Sci.* 2009;**10**:4198-4209.

93. Badadi V Y, Najafi L, Najafi A, Gholami H, Zarji M E B, Golzadeh J, Amraie E and Ali S. **Evaluation of iron oxide NPs effects on tissue and enzymes of liver in rats**. *J Pharmac Biomed Sci.* 2012; **23**:1-4. 8MP primary camera and 2MP front facing camera

94. 11.43 centimeters (4.5-inch) FWVGA capacitive touchscreen with 854 x 480 pixels resolution

95. Android v5.1.1 Lollipop operating system with 1.2GHz Qualcomm 8212 quad core processor, Adreno 302 GPU, 1GB RAM, 8GB internal memory expandable up to 64GB and dual micro-SIM dual-standby (3G+3G) 2070mAH lithium-ion battery 1 year manufacturer warranty for device and 6 months manufacturer warranty for in-box accessories including batteries from the date of purchase

96. Gaharwar U S and Paulraj R. **Iron oxide NPs induced oxidative damage in peripheral Blood cells**. *J Biomedical Sci and Engineering*. 2015; **8**:274-286.

Chapter 5

Palladium (0) Nanoparticles: A Novel and Reusable Catalyst for the Synthesis of Various Pyran Derivatives

Mithu Saha, Amarta Kumar Pal

[1]North Eastern Hill University, Department of Chemistry, Mawlai Campus, Shillong, India

ABSTRACT

A ligand free, Palladium nanoparticles catalyzed synthesis of pyran derivatives using C-H activated compound, malononitrile and aryl aldehyde via Knoevenagel condensation followed by Michael addition reaction using Palladium nanoparticles as catalyst in one-pot is described herein. The advantages of this method lie in its simplicity, low catalyst loading, cost effectiveness and easy to handle. The Palladium Nanoparticles can be reused without loss of activity even after recycling four times. The palladium nanoparticles were characterized by powder X-ray diffraction (XRD), transmission electron microscopy (TEM). The present method also allows us to synthesize highly functionalized title compounds from simple and readily available inputs.

KEYWORDS

Palladium Nanoparticles; Multi-Component Reaction; Pyran Derivatives; Dimedone; 4-Hydroxy Coumarin; Pyranazolone

1. INTRODUCTION

Pyran derivatives are of considerable interest in industry as well as in academia owing to their promising biological activity such as anticoagulant, anticancer, diuretic, spasmolytic and anti-anaphylactin agents. [1] Furthermore, recent studies have revealed several medicinal applications such as, for the treatment of alzheimer's disease, Huntington's disease, amyoprophic lateral sclerosis, Parkinson's diseases, AIDS associated dementia and Down's syndrome as well as for the treatment of schizophrenia and myoclonus. [2,3] These heterocycles are also used as cosmetics, pigments and biodegradable agrochemicals and photoactive materials. [4,5] A considerable effort has been made for the synthesis of pyran annulated heterocyclic derivatives due to their wide applications. Recently, a few methods have been reported by using various catalyst e.g: β-cyclodextrin, [6] TEBA, [7] N-methylmorphine, [8] Et_3N, [9] I_2, [10] organocatalyst, [11] DBU, [12] KF/basic Al_2O_3, [13] ionic liquid, [14] NaBr. [15] Although most of these processes offer distinct advantages, but at the same time they suffer from certain drawbacks also such as longer reaction time, unsatisfactory yields, high costs, harsh reaction conditions, use of stoichiometric amounts of catalyst as well as environmentally toxic catalysts. Thus the development of simple, highly efficient methodologies remains desired.

As a part of our interest in this area, we initiated an investigation to explore the potential of metal nanoparticles for C-C bond formation. The use of palladium nanoparticles has recently received considerable attention in the synthesis of various important heterocycles. [16,17] These metal nanoparticles exhibit size-induced quantumsize effects (i.e., electron confinement and surface effect) and can be exploited for a number of advanced functional applications such as sensors, electronics and catalysis, [18] In general, the catalytic property of metal nanoparticles are a function of their size, crystal lattice parameters. The development of new catalysts in nanorange has emerged as a fertile field for research and innovation. The ability of nanotechnology to enhance catalytic activity opens the potential to replace expensive catalysts with lower amounts of inexpensive nanocatalysts.

Our literature survey at this stage revealed that, there are no reports yet available on the synthesis of pyrans derivatives in one-pot catalyzed by Pd nanoparticles. In the present work, we report here a novel, ligand-free protocol for synthesis of pyran derivatives catalyzed by palladium(0) nanoparticles.

2. EXPERIMENTAL

2.1. Apparatus

General. Melting points were determined in open capillaries and are uncorrected. IR spectra were recorded on Spectrum BX FT-IR, Perkin Elmer (υ_{max} in cm^{-1}) on KBr disks. ^1H NMR and ^{13}C NMR (400 MHz and 100 MHz respectively) spectra were recorded on Bruker Avance II-400 spectrometer in $CDCl_3$ (chemical shifts in δ with TMS as internal standard). Mass spectra were recorded on Waters ZQ-4000. Transmission electron microscope (TEM) was recorded on JEOL JSM 100CX. XRD was recorded on Bruker D8 XRD instrument SWAX. CHN were recorded on CHN-OS analyzer (Perkin Elmer 2400, Series II). Silica gel G (E-mark, India) was used for TLC. Hexane refers to the fraction boiling between 60˚C and 80˚C.

2.2. Synthesis of Pyran Derivatives (6a-k, 7a-e and 8a-i)

A mixture of aryl aldehyde (1 mmol), malononitrile (1 mmol) and activated C-H (3, 4 and 5) (1 mmol) was taken in r.b to that palladium chloride (0.04 mmol), tetrabutylammonim bromide (1 mmol) and sodium carbonate (3 mmol) in acetonitrile (15 mL) was refluxed for the time mention in**Table 1**. After completion of the reaction monitored by TLC the reaction mixture was cooled, filtered and the solvent was removed under reduced pressure. The residue was extracted ethyl acetate (3 × 10 mL) and the combined organic extract was washed with water (3 × 10 mL), brine (10 mL) and dried over anhydrous Na_2SO_4. After removing the solvent the crude product was purified by column chromatography over silica gel (60 - 120 mesh) using hexane-ethyl acetate (7:3) mixture as eluent to afford the pure products (6a-k, 7a-e and 8a-i).

Table 1. Synthesis of pyran derivatives under palladium nanoparticles catalyzed condition.

Entry	Ar-CHO	Product	Time (h)	Yield (%)[a]	m.p (°C)
1	C_6H_5	6a	5	87	230 - 232
2	4-$CH_3C_6H_4$	6b	4.3	88	215 - 217
3	4-$OCH_3C_6H_4$	6c	4.3	85	196 - 198
4	4-FC_6H_4	6d	4	89	189 - 191
5	4-$NO_2C_6H_4$	6e	4	90	174 - 176
6	4-BrC_6H_4	6f	4	88	195 - 197
7	4-$N(CH_3)_2C_6H_4$	6g	5	86	198 - 199
8	4-ClC_6H_4	6h	4.2	88	213 - 215
9	4-CNC_6H_4	6i	4	87	223 - 225
10	2-ClC_6H_4	6j	4.3	89	212 - 213
11	$C_{10}H_{10}$	6k	5	90	210 - 211
12	C_6H_5	7a	4.5	91	255 - 257
13	4-$CH_3C_6H_4$	7b	5	87	256 - 258
14	4-$NO_2C_6H_4$	7c	4	90	250 - 252
15	4-$OCH_3C_6H_4$	7d	4.3	86	227 - 229
16	4-CNC_6H_4	7e	4	91	225 - 226
17	C_6H_5	8a	4	90	167 - 169
18	4-$CH_3C_6H_4$	8b	4.3	90	176 - 178
19	4-FC_6H_4	8c	4	87	170 - 172
20	4-$NO_2C_6H_4$	8d	4	90	190 - 192
21	4-BrC_6H_4	8e	4	88	181 - 183
22	4-ClC_6H_4	8f	4	86	175 - 177
23	4-CNC_6H_4	8g	4.3	89	195 - 197
24	3-$NO_2C_6H_4$	8h	4.1	85	189 - 191
25	2-ClC_6H_4	8i	4.3	88	190 - 192

[a]Isolated yields.

3. RESULT AND DISCUSSION

As a part of our ongoing interest in the synthesis of biologically relevant heterocyclic compound, [19,20] we had the opportunity to further explore its catalytic activity toward the synthesis of pyran derivatives. Recently it has been proved that Pd (0) nanoparticles as catalyst offer great opportunities for a wide range of application in organic synthesis such as Heck reaction, Suzuki-Miyaura coupling, hydrogenation of olefins and Hiyama crosscoupling. [21-24] To optimize the reaction conditions for the synthesis of pyran derivatives (Scheme 1), condensation of dimedone (1 mmol), benzaldehyde (1 mmol) and malononitrile (1 mmol) in presence of palladium nanoparticles in acetonitrile under refluxing condition was selected as a model reaction and it furnished a white solid product 6a in 87% yield. After the completion of the reaction monitored by TLC, the reaction mixture was brought to room temperature, cooled, filtered extracted with ethyl acetate, washed with brine and dried over Na2SO4. The crude mass was purified by column chromatography. Structure of compound 6a was determined by the analysis of analytical and spectral data. The presence of a singlet at δ 4.33 and broad singlet at δ 4.47 in 1H NMR and picks at 3428, 3321, 2196 cm–1 and 1659 cm–1 in IR spectra clearly indicate the formation of 6a.

The Pd (0) nanoparticles were generated in situ by the reduction of $PdCl_2$ in presence of tetra butyl ammonium bromide (TBAB) and sodium carbonate (Na_2CO_3) in acetonitrile under refluxing condition. We assume that, Pd^{2+} was reduced to zero-valent state due to the transfer of two electron from Br^- ion of TBAB. [25-27] The transmission electron microscopy (TEM) (**Figure 1**) micrograph showed that palladium nanoparticles with an average ranging from of 30 - 40 nm. X-ray diffraction (XRD) of palladium nanoparticles are shown in (**Figure 2**).

Furthermore, in transition-metal-catalyzed reactions, ligands play a significant role. Numerous phosphine and nonphosphine ligands for palladium are described in the literature for cross-coupling reactions. Several of these ligands are air and moisture sensitive, difficult to prepare, and expensive. Thus, catalysis by a ligand-free metal center is an area of high importance. [28,29] The catalytic system based on metal nanoparticles appears to be a promising area because of the high surface to volume ratio of such particles.

A plausible mechanism for the reaction was depicted in Scheme 2. Pd Nanoparticles increase the electrophilicity of carbonyl carbon of C-H activated compounds as well as aryl aldehydes. First aryl aldehydes and malononitrile formed Knoevenagel product 3. Methine carbon of 3 is activated by nanoparticles and it react with C-H activated compound in a Michael fashion, giving intermediate 5. Intermediate 5 undergoes intramolecular cyclization followed by tautomerization furnished pyran derivatives 7.

With these encouraging results, we turn to explore the scope of the present protocol with other C-H activated compounds such as 4-hydroxy coumarin or pyrazolone using aryl aldehydes and malononitrile under similar reaction condition (Scheme 1). The components underwent successful condensation to give corresponding pyran derivatives in good to excellent yields. The results are summarized in **Table 1**. From **Table 1**, it is evident that aromatic aldehydes containing either electron donating or electron withdrawing substituent provided good to excellent yields.

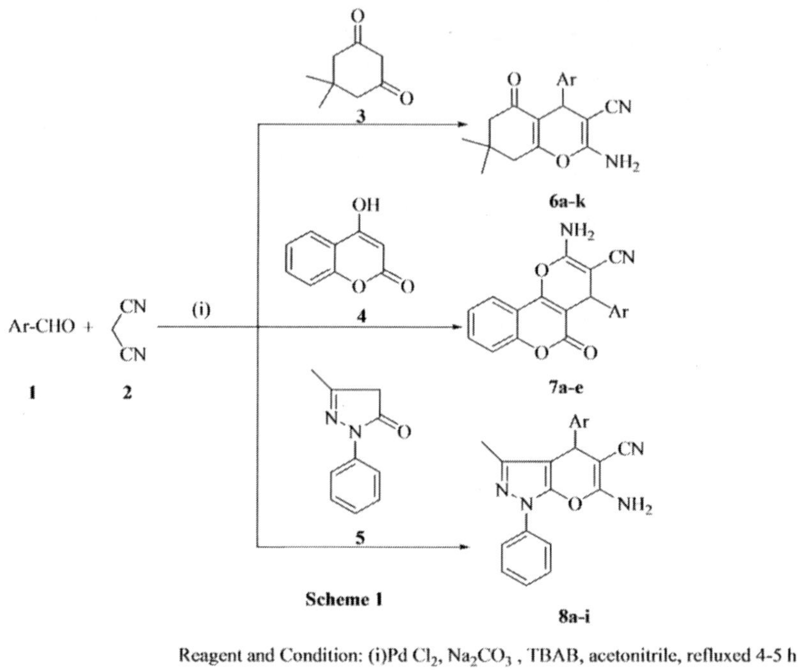

Scheme 1. Synthesis of pyran derivatives by using palladium nanoparticles.

Domino Knoevenagel condensation/Michael addition of dimedone, benzaldehyde and malononitrile using Pd (0) nanoparticles was investigated with various solvents, including water (H_2O), dichloromethane (DCM), and tetrahydrofuran (THF). Significant improvement was achieved in THF and DCM (**Table 2**, entries 1 and 2) but the best result was observed when we used acetonitrile was used as a reaction medium (**Table 2**, entry 5).

We also evaluated the amount of catalyst required for the reaction and we found that yields were obviously affectted by the amount of catalyst loading. We have verified the concentration of catalyst from 0.01 to 0.06 mmol. The result revealed that best yield was obtained by using 0.04 mmol catalyst. With increase the catalyst concentration the yield of the desire product was found to be constant. We also carried out the reaction without any catalyst, but the product was isolated in poor yield (16%), and starting material was recovered. Therefore, the catalyst plays a crucial role in the success of the reaction in term of yields of the product. The results obtained are summarized in**Table 3**.

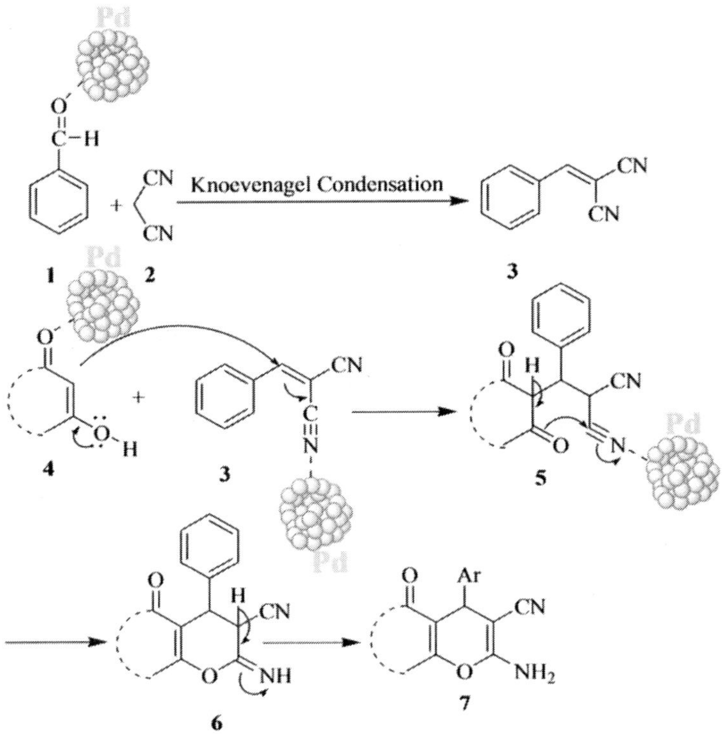

Scheme 2. Plausible reaction mechanism.

For practical application of heterogeneous system, recovery and reusability of the catalyst are very important factors. To clarify this issue, we established a set of experiment using recycled catalyst. The reactions were carried out under similar conditions. After completion of the first set of reaction to afford the corresponding product in 87% yield. The catalyst was recovered by centrifugation of the reaction mixture at 4000 rpm at room temperature for 3 - 5 min and filtered, washed and dried. A new reaction was then performed with fresh solvent and reactants under the same condition. To our satisfaction, palladium catalyst could be used at least four times without any noticeable change in activity (**Figure 3**).

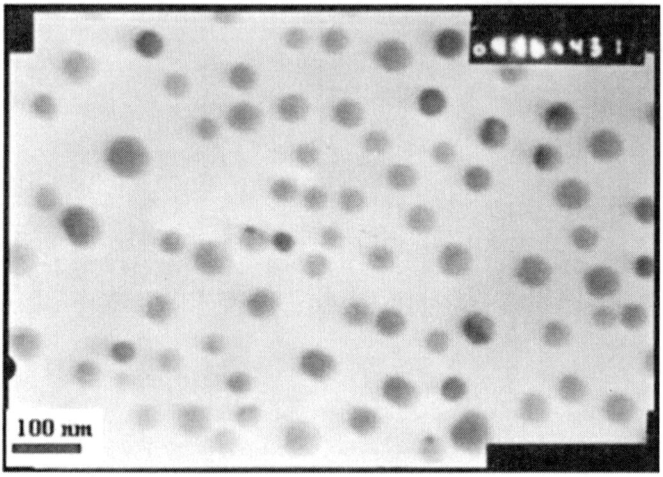

Figure 1. TEM image of palladium nanoparticles.

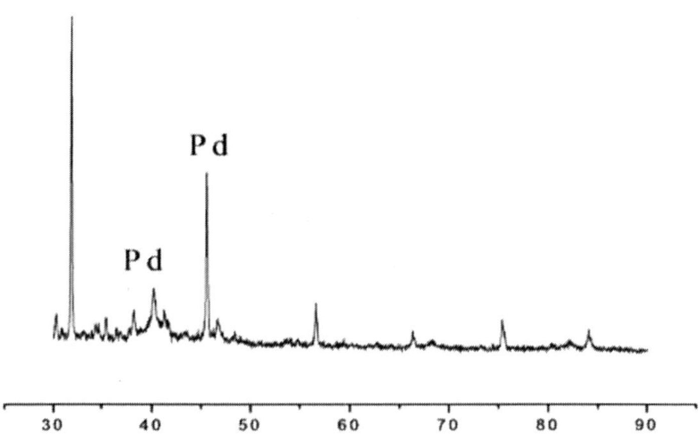

Figure 2. Powder XRD pattern with characteristic dihedral angels at 40.17° and 45.73° corresponding to Pd nanoparticles.

Table 2. Optimization of solvent effect on the model reaction[a].

Entry	Solvent)[b]	Time (h)	Yield (%)[c]
1	Tetrahydrofuran	5	62
2	Dichloromethane	5	67
3	Ethanol	5	53
4	Water	5	51
5	Acetonitrile	5	87

[a]Reaction Conditions: Dimedone (1 mmol), benzaldehyde (1 mmol), malonitrile (1 mmol), and $PdCl_2$ (0.04 mmol), TBAB (1 mmol), Na_2CO_3 (3 mmol), solvent (15 mL). [b]Reaction under refluxing condition. [c]Isolated yields.

Table 3. Effect of catalyst loading on the model reaction[a].

Entry	Catalyst (mmol)	Temperature (°C)	Time (h)	Yield (%)[b]
1	No catalyst	refluxed	5	16
2	0.01	refluxed	5	33
3	0.02	refluxed	5	52
4	0.04	refluxed	5	87
5	0.06	refluxed	5	87

[a]Reaction Conditions: Dimedone (1 mmol), benzaldehyde (1 mmol), malonitrile (1 mmol), and $PdCl_2$ (0.04 mmol), TBAB (1 mmol), Na_2CO_3 (3 mmol), solvent (15 mL). [b]Isolated yields.

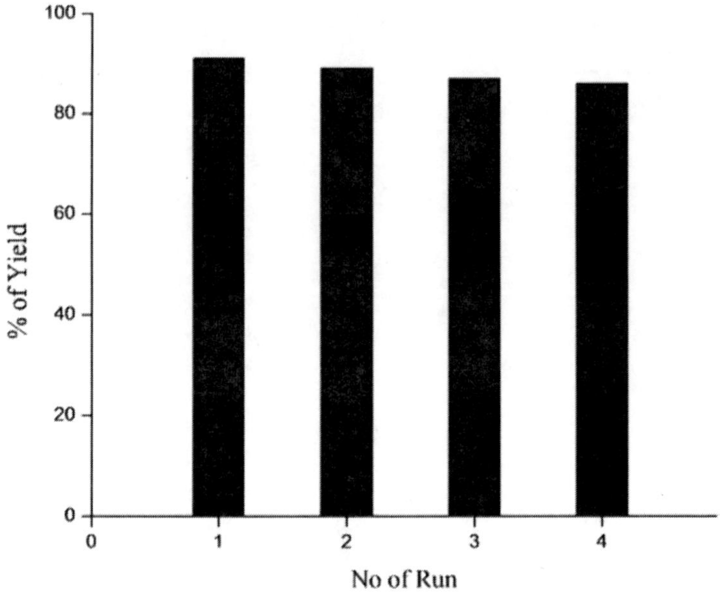

Figure 3. Reusability of the Palladium catalyst.

4. CONCLUSION

In conclusion, we have demonstrated a new and efficient methodology for the preparation of pyran derivatives in one-pot procedure using simple and readily available starting materials. Prominent among the advantages of this new method are novelty, low catalyst loading (0.04 mmol), operational simplicity, environmental friendlyness, good yields and reusability of the catalyst.

ACKNOWLEDGEMENTS

We thank the Department of Chemistry, Sophisticated Analytical and Instrumentation Facility (SAIF) of NorthEastern Hill University and New Delhi for supporting this work under Special Assistance Programme (SAP). We are also thankful to DST for financial support (sanctioned no: SERC/F/0293/2012-13)

REFERENCES

1. L. Bonsignore, G. Loy, D. Secci and A. Calignano, "Synthesis and Pharmacological Activaty of 2-Oxo-(2H)-1- benzopyran-3-carboxamide Derivatives," European Journal of Medicinal Chemistry, Vol. 28, No. 6, 1993, pp. 517-520.

2. W. O. Foye, "Principi Di Chimica farmaceutica," Piccin, Padova, 1991.

3. L. L. Andreani and E. Lapi, "On Some New Esters of Coumarin-3-carboxylic Acid with Balsamic and Bronchodilator Action," Bollettino chimico farmaceutico, Vol. 99, 1960, pp. 583-586.

4. D. Armetso, W. M. Horspool, N. Martin, A. Ramos and C. Seaone, "Synthesis of Cyclobutenes by the Novel Photochemical Ring Contraction of 4-Substituted-2-amino-3,5- dicyano-6-phenyl-4H-pyrans," The Journal of Organic Chemistry, Vol. 54, No. 11, 1989, pp. 3069-3072.

5. E. A. A. Hafez, M. H. Elnagdi, A. G. A. Elagamey and F. M. A. A. Ei-Taweel, "Nitriles in Heterocyclic Synthesis: Novel Synthesis of Benzo[c]coumarin and of Benzo[c] pyrano[3,2-c]quinoline Derivatives," Heterocycles, Vol. 26, 1987, pp. 903-907.

6. K. Kanagaraj and K. Pitchumani, "Solvent-Free Multicomponent Synthesis of Pyranopyrazoles: Per-6-amino-bcyclodextrin as a Remarkable Catalyst and Host," Tetrahedron Letters, Vol. 51, No. 25, 2010, pp. 3312- 3316.

7. D. Shi, J. Mou, Q. Zhuang, L. Niu, N, Wu and X. Wang, "Three-Component One-Pot Synthesis of 1,4-Dihydropyrano[2,3-c]pyrazole Derivatives in Aqueous Media," Synthetic Communications, Vol. 34, No. 24, 2004, pp. 4557- 4563.

8. F. Lehmann, M. Holm and S. Laufer, "Three-Component Combinatorial Synthesis of Novel Dihydropyrano[2,3-c] pyrazoles," Journal of Combinatorial Chemistry, Vol. 10, No. 3, 2008, pp. 364-367.

9. G. Vasuki, K. Kumaravel, "Rapid Four-Component Reactions in Water: Synthesis of Pyranopyrazoles," Tetrahedron Letters, Vol.49, No. 39, 2008, pp. 5636-5638.

10. R. S. Bhosale, C. V. Magar, K. S. Solanke, S. B. Mane, S. S. Choudhary and R. P. Pawar, "Molecular Iodine: An Efficient Catalyst for the Synthesis of Tetrahydrobenzo[b] pyrans," Synthetic Communications, Vol. 37, No. 24, 2007, pp. 4353- 4357.

11. S. Muramulla and C.-G. Zhao, "A New Catalytic Mode of the Modularly Designed Organocatalysts (MDOs): Enantioselective Synthesis of Dihydropyrano[2,3-c]pyrazole," Tetrahedron Letters, Vol. 52, No. 30, 2011, pp. 3905- 3908.

12. J. M. Khurana, B. Nand and P. Saluja, "DBU: A Highly Efficient Catalyst for One-Pot Synthesis of Substituted 3,4-Dihydropyrano[3,2-c]chromenes, Dihydropyrano[4,3- b]pyranes, 2-Amino-4hbenzo[h]chromenes and 2-Amino- 4H Benzo[g]chromenes in Aqueous Medium," Tetrahedron Letters, Vol. 66, 2010, pp. 5637-5641.

13. J.-T. Li, W.-Z. Xu, L.-C. Yang and T.-S. Li, "One-Pot Synthesis of 2-Amino-4-aryl-3-carbalkoxy-7,7-dimethyl- 5,6,7,8-tetrahydrobenzo[b]pyran Derivatives Catalyzed by KF/Basic Al_2O_3 under Ultrasound Irradiation," Synthetic Communications, Vol. 34, No. 24, 2004, pp. 4565- 4571.

14. X.-S. Wang, M.-M. Zhang, H. Jiang, C.-S. Yao and S.-J. Tu, "Three-Component Green Synthesis of N-Arylquinoline Derivativesin Ionic Liquid [BmimD][BF_4L]: Reactions of Arylaldehyde,3-arylamino-5,5-dimethylcyclohex- 2-enone, and Activemethylene Compounds," Tetrahedron, Vol. 63, No. 21, 2007, pp. 4439-4449.

15. I. Devi and P. J. Bhuyan, "Sodium Bromide Catalysed One-Pot Synthesis of Tetrahydrobenzo[B]Pyrans via a Three-Component Cyclocondensation under Microwave Irradiation and Solvent Free Conditions," Tetrahedron Letters, Vol. 45, No. 47, 2004, pp. 8625-8627.

16. V. Calo, A. Nacci, A. Monopoli and F. Montingelli, "Pd Nanoparticles as Efficient Catalysts for Suzuki and Stille Coupling Reactions of Aryl Halides in Ionic Liquids," The Journal of Organic Chemistry, Vol. 70, 2005, pp. 6040-6044.

17. B. Baruwati, D. Guin and S. V. Manorama, "Pd on Surface-Modified $NiFe_2O_4$Nanoparticles: A Magnetically Recoverable Catalyst for Suzuki and Heck Reactions," Organic Letters, Vol. 9, No. 26, 2007, pp. 5377-5380.

18. J.-U. Park, J. Heon Lee, U. Paik, Y. Lu and J. A. Rogers, "Nanoscale Patterns of OligonucleotidesFormed by Electrohydrodynamic Jet Printing with Applications in Biosensing and Nanomaterials Assembly," Nano Letters, Vol. 8, No. 12, 2008, pp. 4210-4216.

19. M. Saha and A. K. Pal, "Palladium(0) Nanoparticles: An Efficient Catalyst for the One-Pot Synthesis of Polyhydroquinolines," Tetrahedron Letters, Vol. 52, No. 38, 2011, pp. 4872-4877.

20. M. Saha, S. A. K. Pal and S. Nandi, "Pd (0) NPs: A Novel and Reusable Catalyst for the Synthesis of bis(heterocyclyl) methanes in Water," RSC Advance, Vol. 2, No. 16, 2012, pp. 6397-6400.

21. T. M. Razler, Y. Hsiao, F. Qian, R. Fu, R. K. Khan and W. Doubleday, "A Preparatively Convenient Ligand-Free Catalytic PEG 2000 Suzuki-Miyaura Coupling" The Journal of Organic Chemistry, Vol. 74, No. 3, 2009, pp. 1381-1384.

22. Z. Zhang, Z. Zha, C. Gan, C. Pan, Y. Zhou, Z. Wang and M.-M. Zhou, "Catalysis and Regioselectivity of the Aqueous Heck Reaction by Pd(0) Nanoparticles under Ultrasonic Irradiation," The Journal of Organic Chemistry, Vol. 71, No. 11, 2006, pp. 4339-4342.

23. N. Semagina, A. Renken, D. Laub and L. Kiwi-Minsker, "Synthesis of Monodispersed Palladium Nanoparticles to Study Structure Sensitivity of Solvent-Free Selective Hydrogenation of 2-Methyl-3-butyn-2-ol," Journal of Catalysis, Vol. 246, No. 2, 2007, pp. 308-314.

24. D. Srimani, A. Bej and A. Sarkar, "Palladium Nanoparticle Catalyzed Hiyama Coupling Reaction of Benzyl Halides," The Journal of Organic Chemistry, Vol. 75, No. 12, pp. 4296-4299.

25. M. T. Reetz and M. Maase, "Redox-Controlled SizeSelective Fabrication of Nanostructured Transition Metal Colloids," Advanced Materials, Vol. 11, No. 9, 1999, pp. 773-777.

26. M. T. Reetz and E. Westermann, "Phosphane-Free Palladium-Catalyzed Coupling Reactions: The Decisive Role of Pd Nanoparticles," Angewandte Chemie International Edition, Vol. 39, No. 1, 2000, pp. 165-168.

27. L. Adak, K. Chattopadhyay and B. C. Ranu, "Palladium Nanoparticle-Catalyzed C-N Bond Formation. A Highly Regioand Stereoselective

Allylic Amination by Allyl Acetates," The Journal of Organic Chemistry, Vol. 74, No. 10, 2009, pp. 3982-3985.

28. R. M. Crooks, M. Zhao, L. Sun, V. Chechik and L. K. Yeung, "Dendrimer-Encapsulated Metal Nanoparticles: Synthesis, Characterization, and Applications to Catalysis," Accounts of Chemical Research, Vol. 34, No. 3, 2001, pp. 181-190.

29. M. Moreno-Manas and R. Pleixats, "Formation of Carbon-Carbon Bonds under Catalysis by TransitionMetal Nanoparticles," Accounts of Chemical Research, Vol. 36, No. 8, 2003, pp. 638-643.

Chapter 6

Green Synthesis of Gold Nanoparticles Using Glycerol as a Reducing Agent

Pradnya Nalawade, Tulsi Mukherjee, Sudhir Kapoor*

Radiation & Photochemistry Division, Bhabha Atomic Research Centre, Mumbai, India

ABSTRACT

We report one pot synthesis of uniform and stable polyvinyl pyrolidone (PVP) protected gold nanoparticles (Au NPs) using environmental friendly, glycerol as reducing agent. The effect of the presence of a capping agent (PVP) and the concentration of reactants (glycerol, tetra chloroauric acid, and NaOH) on the size and homogeneity of the Au NPs formed were investigated. Highly stable and well-distributed Au NPs were obtained at higher concentration of NaOH in the presence of PVP with a clear dependence of the size and the concentration of glycerol, NaOH and the presence of capping agent, whereas, large heterogeneous Au NPs were obtained in absence of PVP. The particle morphology, size and crystallinity were characterized using UV-Vis spectroscopy, transmission electron microscopy and X-ray diffraction techniques. The catalytic performance of as synthesized Au NPs for the reduction of o-nitro aniline was investigated in aqueous solution. The pseudo-first-order rate constants were also calculated for the catalytic reaction.

KEYWORDS

Gold Nanoparticles; Glycerol; Synthesis; Catalytic Reaction; Characterization

1. INTRODUCTION

In recent decades there in growing interest in the synthesis of metal nanoparticles due to their unique properties, which are significantly different from the behavior of the respective bulk material [1,2]. Gold nanoparticles (Au NPs) have wide attraction because of their electronic, biosensing, plasmonic, photonics, catalysis, biomedical and surface-enhanced Raman scattering (SERS) properties [2-7]. Most of the chemical methods reported in the literature for the synthesis of Au NPs often involve use of toxic reducing agents (such as sodium borohydride, hydrazine, etc.) and harsh reaction parameters like high temperatures in the polyol method [8-10]. Glycerol, also known as glycerin, is commonly used in the pharmacological application and its derivative used in the synthesis of drugs [11]. It is also used as a sweetener in food industry and due to its hygroscopic nature it is widely used as dehydrating and moistening agent [12]. Besides that it is easily biodegradable in aerobic conditions thus can be replaced by traditional reducing agent.

Few studies have been reported in the literature demonstrating the formation of silver and Au NPs using glycerol as reducing agent at low temperature. Genc et al. have shown low temperature method to obtain monodispersed Au NPs using glycerol-incorporated nanosized liposome, where glycerol, is incorporated on both the external and internal polar surfaces of liposome encapsulating chloroauric acid, facilitates the reduction of Au(III) to form Au(0) atoms and subsequently nanoparticles [13]. Singh et al. have reported the formation of nickel nanoparticles in glycerol at 100°C by using hydrazine hydrate in alkaline medium [14]. Nisaratanaporn and Wongsuwan prepared silver powders of particle size more than 63 nm using silver alkoxide as silver ion precursor and glycerol as a reducing agent at high temperature [15]. Grace and Pandian reported the synthesis of spherical and prism shaped Au NPs in glycerol at both reflux and microwave conditions [16].

As water is the most beneficial solvent it will be of interest to study the effect of water in glycerol on the formation of NPs. To the best of our knowledge, there is no report demonstrating the effect of concentration of glycerol, NaOH and capping agent on the formation of Au NPs at room temperature without requiring any additional reactants. In this paper we report environmentally friendly, low temperature method to prepare Au NPs without using any external reducing agent. The formed particles were further studied for their catalytic application in reduction of o-nitro aniline.

2. MATERIALS AND METHODS

2.1. Materials

Tetra chloroauric acid, glycerol, PVP (Mol. wt. 40000), sodium hydroxide, o-nitro aniline and sodium borohydride were purchased from Sigma-Aldrich and used as received. Millipore purified water having 18.2 MΩ electrical resistivity was used for making solutions.

2.2. Characterization

Absorption measurements were carried out on a Jasco V- 650 spectrophotometer. The spectra were recorded at room temperature using 1 cm quartz cuvette. Samples for transmission electron microscopy (TEM) were prepared by putting a drop of the colloidal solution on a copper grid coated with a thin amorphous carbon film placed on a filter paper. The excess solvent was removed using a filter paper, and letting the solvent evaporate at room temperature. TEM characterization was carried out using a Phillips CM 200 electron microscope with working voltage 200 kV having magnification 34,000 to 78,000. Particle size was measured by TEM photograph and calculated the size by considering at least 100 particles. XRD measurement was carried out on precipitated nanoparticles using Philips X'pert Pro machine with monochromatised CuKα X-ray source operated at 20 kV and 30 mA.

2.3. Method

2.3.1. Synthesis of Au NPs in Glycerol or Water:

Glycerol System

All the solutions were prepared freshly in order to avoid any photochemical reactions and experiments were carried out in an aerated condition. Gold sol was prepared by mixing the required concentration of NaOH (5×10^{-4} M - 1×10^{-3} M) and chloroauric acid solution (1 × 10−4 M - 1 × 10−3 M) in either neat glycerol or glycerol-water mixtures [80 to 20% (v/v) glycerol in water] at room temperature in the absence or presence of PVP [0.05% - 0.1% (w/v)]. Gradual formation of different color with time indicates the formation of Au NPs (Scheme 1). The experiments were repeated at least three times and found to be within the experimental error of ± 5%.

2.3.2. Catalytic Reduction of o-Nitro Aniline

The catalytic reduction reaction was carried out in 5 ml volumetric flask. In a typical reaction excess of ice cold $NaBH_4$ (3×10^{-2} M) was mixed with o-nitro aniline (2.3 × 10−4 M) solution in water. This was followed by the addition of Au NPs. The absorption spectra were recorded immediately after mixing with a time interval of 2 min in a scanning range of 200 - 600 nm at 25˚C till the yellow colored o-nitro aniline solution became colorless (from 4 - 25 min depending upon the concentration of catalyst used). The experiments were repeated at least three times and found to be within the experimental error of ± 10%.

3. RESULTS AND DISCUSSION

3.1. Formation Au NPs in Neat Glycerol

Metal nanoparticles (especially Ag, Au and Cu) exhibit a unique UV-Vis absorption band derived from collective oscillation of conduction electrons upon interaction with electromagnetic radiation, known as surface plasmon resonance (SPR) [2]. The absorption maximum of SPR depends on the shape and size of the particles [2]. This optical property of noble metal NPs is exploited for many applications [2]. In the present work Au NPs were prepared by simply adding NaOH to the tetra chloroauric acid solution in neat glycerol. Formation of nanoparticles was observed, after 2 min of shaking, by change in the color of the solution from light yellow to colorless to pinkish violet. **Figure 1** shows the typical changes in absorption UV-Vis spectra of the Au NPs in neat glycerol. The initial peaks at 505 nm due to the surface plasmon resonance of Au NPs which is red shifted up to 536 nm along with formation of new band at longer wavelength (748 nm) with time, this might be due to formation of the anisotropic Au NPs which always exhibit two to three SPR bands depending on their shape compared with single SPR for small spherical nanocrystals [17]. As can be seen from the TEM image of the Au NPs, prepared in neat glycerol (**Figure 1**(b)), that the formed particles are nonspherical including nanorod with average particle size 30 nm. This non spherical nature of the nanoparticles caused the absorption at longer wavelength in absence of stabilizing agent. Yang et al. have shown the formation of anisotropic Au NPs of average diameter of 50 - 100 nm without using any stabilizing agent [18]. The colloidal stability of particles prepared in neat glycerol was not good as they starts to aggregate after about 24 hrs. **Figure 2** shows the XRD pattern in the 2q range 30° - 70° for Au NPs prepared in neat glycerol. The patterns exhibit peaks due to diffraction from (111), (200) and (220) planes of metallic gold (JCPDS No. 04-0784). The XRD pattern suggests the formation of crystalline Au NPs with face-centred cubic (fcc) structure. Broadening of XRD peaks clearly indicate that the samples are nanocrystalline in nature. The particle size (d) was also calculated from XRD line broadening data, after instrumental line broadening correction, using Scherrer formula, $k\lambda = d.\beta.\cos\theta$, where λ is wavelength of the X-ray used, and it is 1.5406 Å, β is the angular line width at half maximum intensity, θ is Bragg diffraction angle and k is a constant, and is equal to 0.9 [19]. The particle size calculated from**Figure 2**, was found to be ~25 nm.

3.2. Effect of PVP Concentration

Figure 3 also shows the effect of concentrations of PVP on UV-Vis optical absorption spectra of the Au NPs sols prepared by addition of 1×10^{-3} M NaOH and 5×10^{-4} M tetra chloroauric acid in neat glycerol. PVP was found to be a very efficient stabilizer for the stabilization of Au NPs. Even a small concentration of PVP (0.05%) was sufficient to stabilize the Au NPs. When concentration of PVP decreased to 0.01% the red shift of absorption maxima was observed which might be due to the increase in particle size of the Au NPs. However, concentration of PVP (^{3}0.05%) has no significant effect on stability as well as size, as Au NPs prepared with different concentrations are equally stable for more than a year. Since the particles were stable in the sol form XRD could not be carried out, nevertheless, TEM and selective area electron diffraction (SAED) confirmed the presence of fcc Au NPs (results not shown).

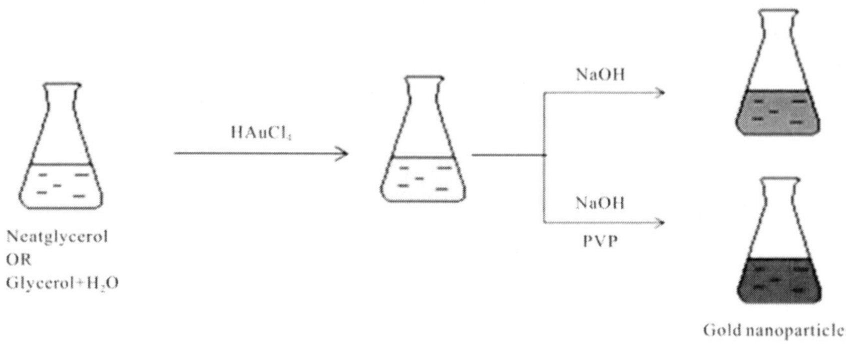

Scheme 1. Synthesis of gold nanoparticles.

3.3. Effect of NaOH Concentration

The above results showed that concentration of PVP (^{3} 0.05%) had no significant effect on the stability of the particles; hence, for the preparation of particles we had selected 0.1% PVP to make sure that the particles do not aggregate. **Figure 4** shows the effect of NaOH concentration (5×10^{-4} M - 1×10^{-3} M) on the UV-Vis absorp- tion spectrum of Au NPs in presence of 0.1% PVP in neat glycerol. The blue shift of SPR band with increasing concentration of NaOH indicates the decrease in particle size of Au NPs. The corresponding TEM image of Au NPs formed in presence of 5×10^{-4} M - 1×10^{-3} M NaOH (Figures 5(a) and (b)) shows that the average particle size falls from 20 nm to 5 - 7 nm when the NaOH concentration was increased from $5 \times 10^{-4\ M\ -\ 1\ \times}$ 10–3 M NaOH. Since the particles were stable in the sol form XRD could not be carried out. Chen et al. have observed similar results where palladium nanoparticles in the size range of 8.6 to 2.4 nm were prepared by changing the concentration of NaOH in presence of PVP [20]. However, in our case, no particle formation was observed when concentration of NaOH was below 5 × 10–4 M.

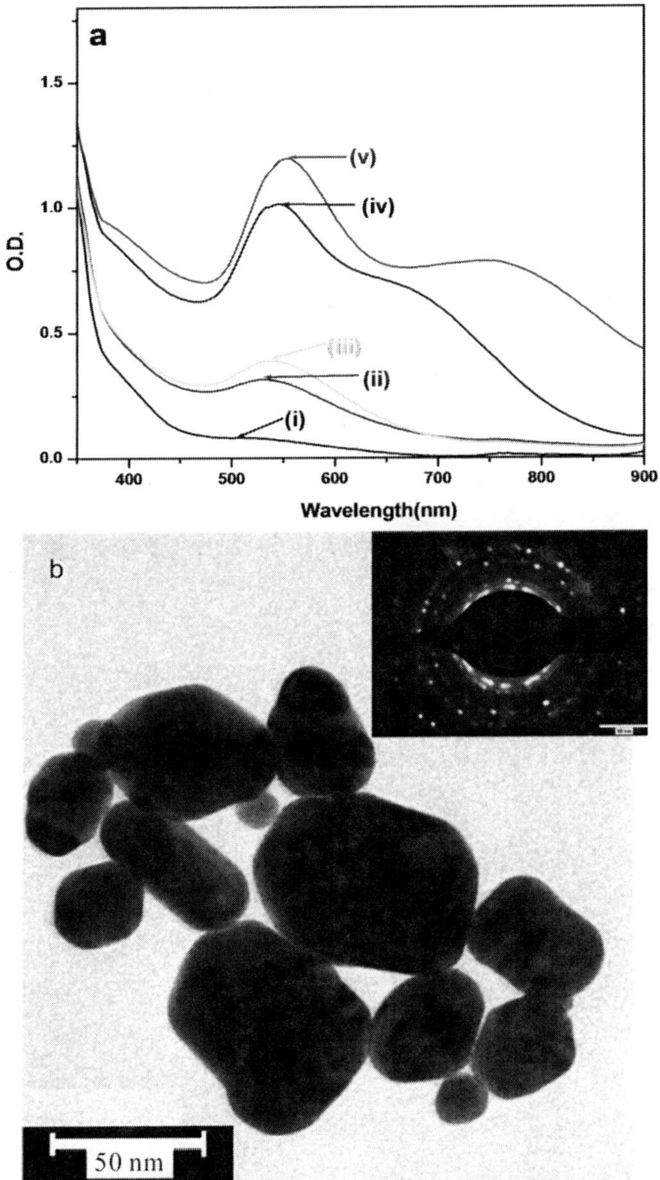

Figure 1. (a) Variation in the surface plasmon absorption band of Au NPs formed on mixing 5×10^{-4} M $HAuCl_4$ and 1×10^{-3} M NaOH in neat glycerol (i) after 3 min (ii) after 5 min (iii) after 48 min (iv) after 24 hr (v) after 6 days; (b) Corresponding TEM image and electron diffraction pattern.

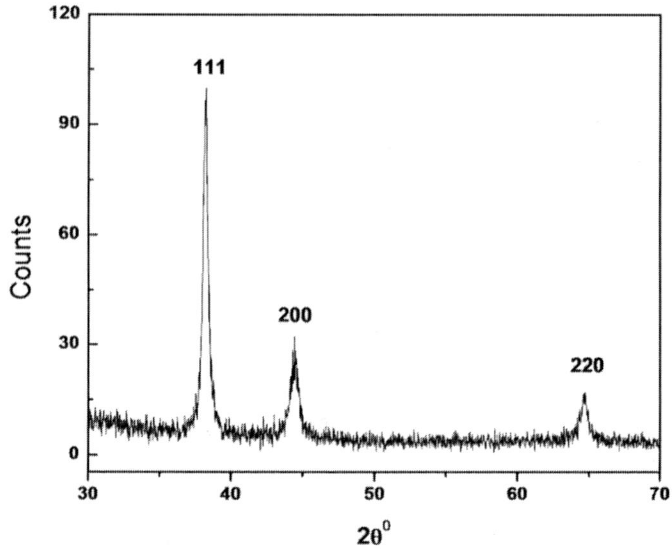

Figure 2. XRD pattern of Au NPs.

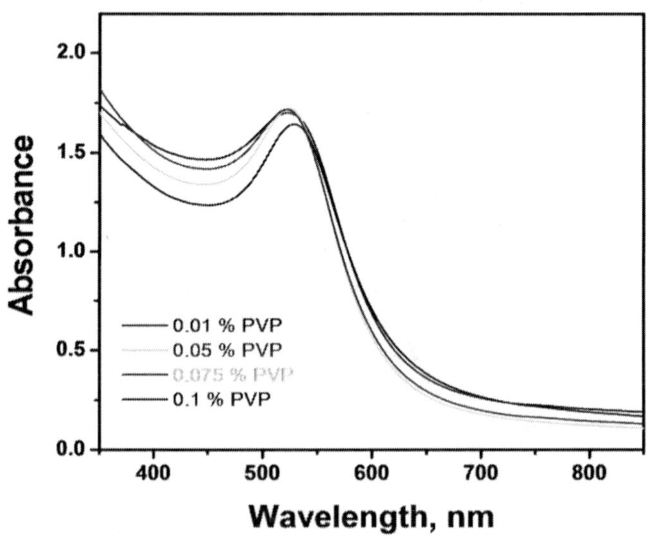

Figure 3. Effect of PVP concentration (wt %) on the optical absorption spectrum of Au NPs prepared by addition of 1×10^{-3} M NaOH and of 5×10^{-4} M tetra chloroauric acid. (i) 0.01 (ii) 0.05 (iii) 0.075 and (iv) 0.1. Spectra were recorded after 48 hr.

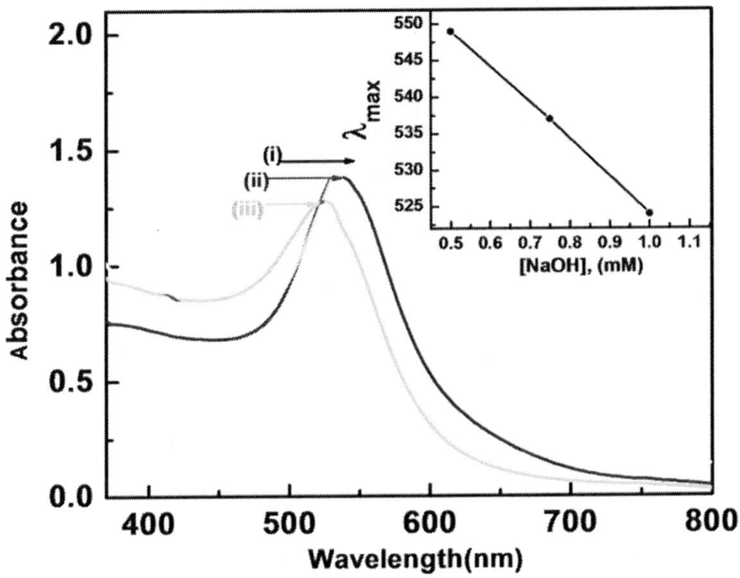

Figure 4. Effect of NaOH concentration on Au NPs sol synthesized by the reaction of (i) 5×10^{-4}, (ii) 7.5×10^{-4} and (iii) 1×10^{-3} M NaOH with 5×10^{-4} M auric acid in the presence of 0.1% PVP in glycerol.

3.4. Effect of Gold Ion Concentration

On addition of NaOH (1×10^{-3} M) to different concentrations of the tetra chloroauric acid solution in absence of PVP, the color of the solution changed from light yellow to colorless to shades of blue and violet. However, in presence of PVP the color of the gold sol range from wine red to pink to magenta to violet for gold ion concentration of 1×10^{-4} M, 2×10^{-4} M, 5×10^{-4} M and 7.5×10^{-4} M, respectively. **Figure 6** shows the UV-Vis spectra of the Au NPs obtained for various concentrations of gold ions in the absence and presence of 0.1% PVP, at room temperature. The peak at ~530 nm corresponds to the surface plasmon absorption of Au NPs. The absorption peak intensity of Au NPs in both absence and presence of PVP was found to increase and the l_{max}gradually shifted towards higher wavelength with the increasing concentration of gold ions, this is possibly due to the formation of large size spherical nanoparticles at higher concentrations of gold ions. However, for the gold ions concentration of 5×10^{-4} M and above in absence of PVP,

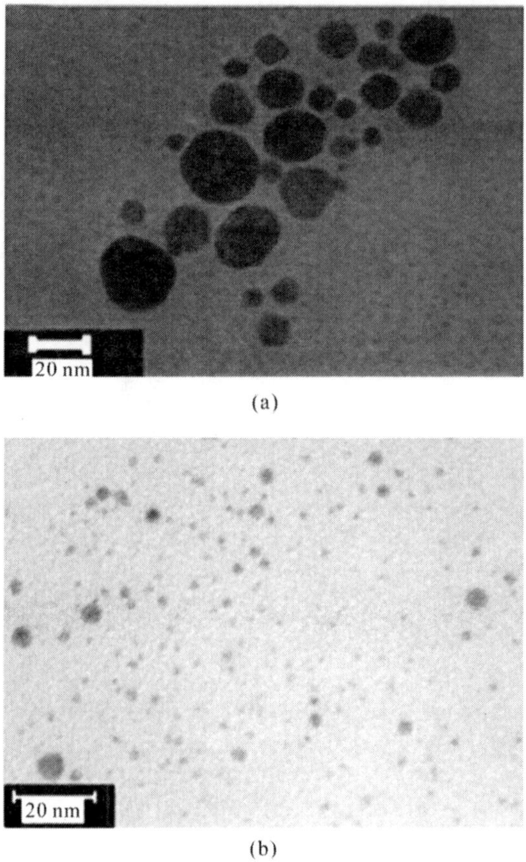

Figure 5. TEM image of Au NPs obtained with different concentration of NaOH other conditions are same as **Figure 4**. (a) 7.5×10^{-4} M (b) 1×10^{-3} M.

the shoulder at 635 nm in the case of 5×10^{-4} M or new peak at 735 nm in the case of 7.5×10^{-4}M was observed which is possibly due to formation of nonsherical Au NPs as explained in **Figure 1**. Similar results were obtained by S. Boopathi et al. in the synthesis of Au NPs using tetraphenylborate and PVP for colorimetric detection of mercury ions [21]. Oza et al. studied the effect of chloroauric acid concentration on the formation of gold nanorods in presence of cetyl trimethylammonium bromide (CTAB) and it was found that high yield of nanorods formation occurred when the concentration of gold ions was $\geq 1 \times 10^{-3}$ M, whereas, yield of spherical Au NPs was high at lower concentration of gold ions [22]. However, in our study it was found that, on addition of stabilizing agent PVP during the synthesis, the band at longer wavelength was not observed indicating formation of almost spherical nanoparticles. This might be due to the combine effect of viscosity of medium and adsorption of PVP on the surface of nanoparticles during growth which result in to the formation of spherical nanoparticles. The increased absorbance may be correlated to the production of metal nanoparticles in larger quantities in the presence of higher concentration of gold ion precursor. It was also seen that at concentration of 5 × 10^{-4} M and 7 × 10^{-4} M tetra chloroauric acid more stable, spherical and small size Au NPs were formed. Also, the yield of the Au NPs was reasonably good. At both the concentrations similar size of particles (~10 nm) were observed (results not shown). As mentioned above due to the good stability of the particles XRD could not be carried out.

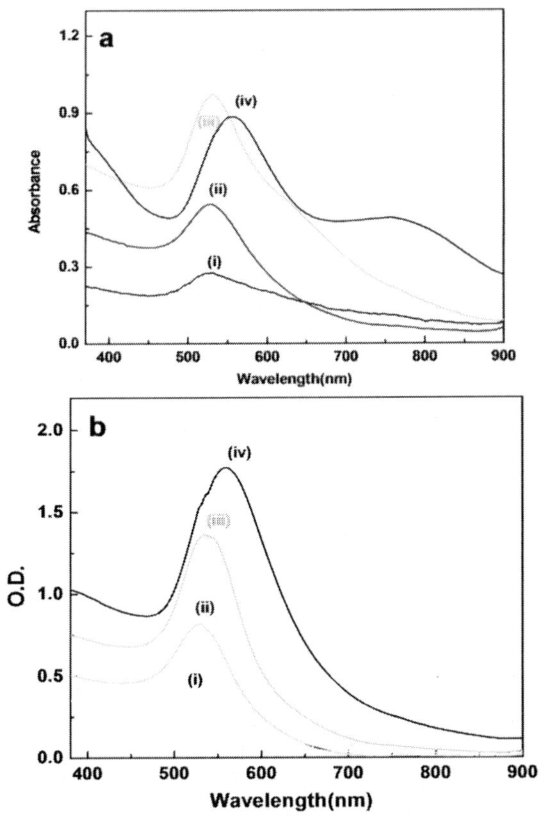

Figure 6. UV-Vis spectra showing effect of tetra chloroauric acid concentration on Au NPs sol synthesized by the reaction of (i) 1 × 10^{-4}, (ii) 2 × 10^{-4}, (iii) 5 × 10^{-4}and (iii) 7.5 × 10^{-4}M auric acid with 5 × 10^{-4} M NaOH in glycerol. (a) without PVP (b) with 0.1 % PVP. All spectra were taken after 24 hr.

Based on above results to get the good dispersion as well as stable Au NPs in neat glycerol the optimum concentrations of tetra chloroauric acid, NaOH and PVP are 5×10^{-4} M, 1×10^{-3} M and 0.1 % (w/v), respectively.

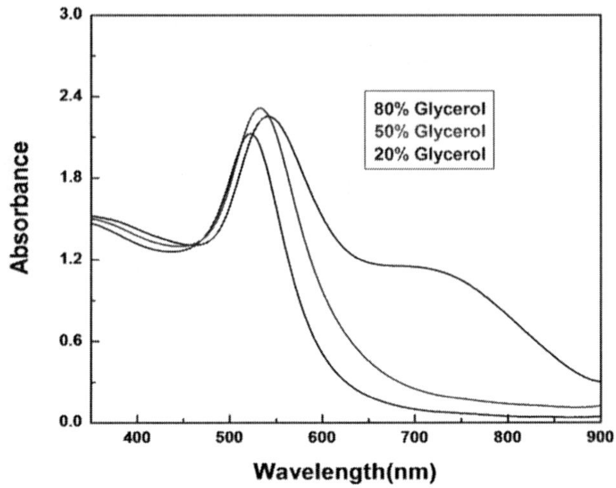

Figure 7. Effect of glycerol concentration on the UV-Vis spectra of Au NPs in presence of 1×10^{-3} M NaOH and 7×10^{-4} M $HAuCl_4$ without using PVP.

3.5. Synthesis of Au NPs in Glycerol:

Water Mixture

To see the effect of concentration of glycerol on the particle size and shape of nanoparticles, synthesis of Au NPs were carried out at different concentrations of glycerol in water with fixed concentration of NaOH without using PVP. **Figure 7** shows the UV-Vis spectra of the Au NPs obtained with different concentration (v/v) of glycerol. In 80% glycerol two SPR band centered at 543 and 694 due of Au NPs was observed. This new band formation at longer wavelength indicates the presence of nonspherical nanoparticles along with spherical NPs. However when concentration of glycerol decreased up to 20%, almost spherical and monodispersed nanoparticles were obtained.**Figure 8** shows the TEM images of Au NPs obtained at 80%, 50% and 20 % of glycerol. The shape and average particle size obtained by TEM measurements at different concentration of glycerol is summarized in **Table 1**.

The above results show that at low concentrations of glycerol it is possible to stabilize the particles without using stabilizer. It seems that the stability of the particles depend on the viscosity of the medium though not in direct proportionality. High viscosity means that formation of nuclei in a smaller zone and thus the interaction of particles in a smaller confined volume results in aggregation of the particles. At appropriate viscosity the interaction between the particles decreases and this leads to the stability of the particles. Thus, it can be concluded that for the preparation of small and stable Au NPs the mixture of glycerol:water is better as compared to neat glycerol.

3.6. Catalytic activity of Au NPs

Aromatic amino compounds are widely used in industry as an intermediate in the preparation of polymers, azo dyes, etc. However, the chemical reduction of aromatic nitro compounds with sodium borohydride is extremely slow, and the use of a catalyst is essential. Of late, metal nanoparticles have been explored for this reaction. In this the reduction of nitro aromatics to its corresponding amino derivatives with an excess amount of $NaBH_4$ has frequently been used as a model reaction to examine the catalytic performance of metal nanoparticles [23,24].

It is often discussed in the literature that the presence of stabilizer can inhibit the catalytic activity of the metal nanoparticles [2]. In the present work it was possible to stabilize small Au NPs in the absence of PVP in 20:80 (v/v) glycerol:water mixture. Hence, for catalytic reaction Au NPs prepared in 20:80 (v/v) glycerol:water mixture were used; however, the catalytic reaction was carried out in an aqueous solution. The strong UV-Vis peaks characteristic of o-nitro aniline appears at 412 nm and 283 nm [25]. It was found that there was a very slow decrease of absorbance during the chemical reaction without catalyst. However, after the addition of Au nanoparticles a decrease in the absorbance at 412 nm and shift of the peak from 283 nm to 289 nm with reaction time usually indicate steady reduction of o-nitro aniline to benzenediamine (**Figure 9**). The progress of reduction reaction was monitored by measuring UV-Vis absorption recorded at different times; t, at 298 K at different concentration of Au NPs. The ratio of C_t and C_0, where C_t and C_0 refer to the concentration of o-nitro aniline at times t and 0, is measured from the relative intensity ratio of the respective absorbance, at 412 nm. The linear correlation of $\ln(C_t/C_0)$ versus time is shown in inset of **Figure 9** indicating the reaction follows pseudo first-order kinetics. The rate constant (k) for the reaction, directly calculated from the slope of the straight line, for different concentration of Au NPs are tabulated in **Table 2**. It was found that when the gold content of the dispersion (ranging from 14 to 82 mg/mL) was varied keeping other parameters constant, to determine the effect of the amount of catalyst on the rate of the reaction, the rate constant showed an increase of factor ca. 10 times.

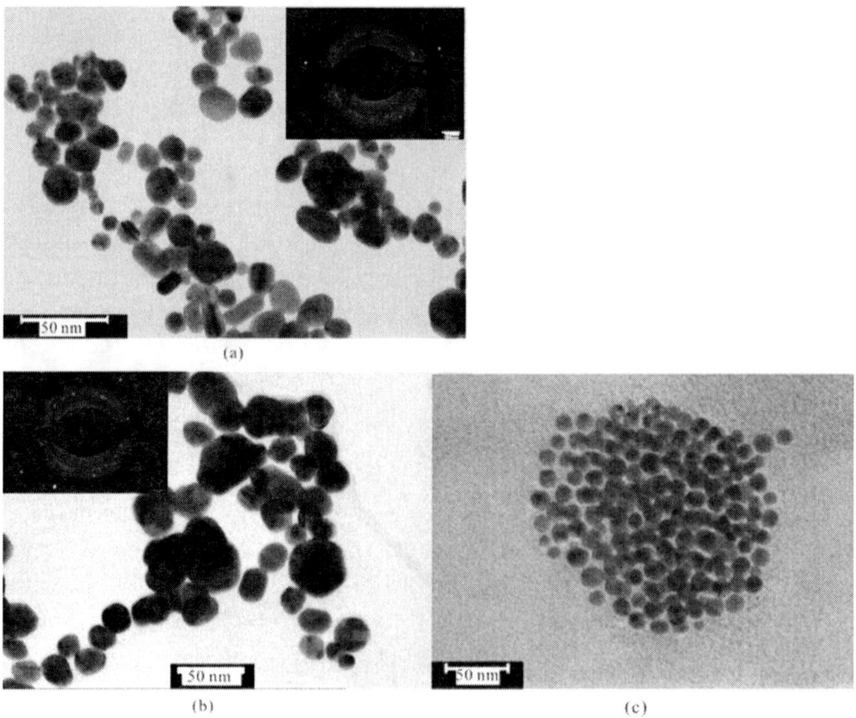

Figure 8. TEM images of Au NPs obtained in presence of different concentration of glycerol (a) 80% (b) 60% and (c) 20%.

Table 1. Summary of particle size and shape of Au NPs at different glycerol:water ratios (v/v) in the absence of PVP.

% conc. of glycerol	SPR peak (nm)	Shape and size distribution	Particle size by TEM (nm)	Colloidal stability
100%	554 748	Non spherical, Polygonal and polydispersed	50	Turbid after 1 day
80%	543 694	Non spherical, polygonal and polydispersed	25	Stable at least up to 8 months
50%	533	Spherical and nearly monodispersed	20	Stable at least up to 8 months
20%	522	Spherical and monodispersed	8	Stable at least up to 8 months

Table 2. Variation in the rate of reduction of o-nitro aniline with increasing amount of Au NPs.

Sr. No.	Concentration of Au NPs (μg/mL)	k_{obs}, s^{-1}
1	14	2.5×10^{-3}
2	27	3.2×10^{-3}
3	54	7.5×10^{-3}
4	82	3×10^{-2}

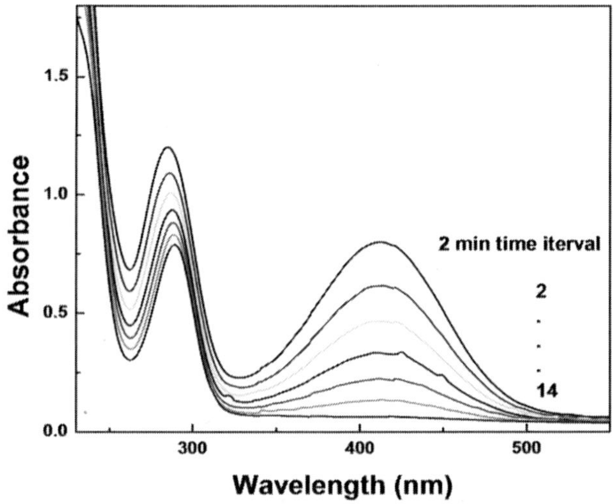

Figure 9. Time-dependent variation in the absorption spectrum of o-nitro aniline reduced by 5 μL/mL of Au NPs prepared in 20% glycerol. Inset: Plot of $\ln(C_t/C_0)$ versus time.

4. CONCLUSION

Green synthesis of Au NPs was carried out by simply adding NaOH to the tetra chloroauric acid solution prepared in glycerol. Glycerol is used both as a reducing and stabilizing agent. The different size and shape of Au NPs can be prepared by varying the concentration of glycerol, concentration of NaOH and concentration of gold ions. Generally, stabilizers are added in the synthesis of the nanoparticles to prevent the aggregation of particles. The results of the present study show that the viscosity of the medium can play a similar role as of stabilizer by decreasing the interaction of the particles. However, there is no direct relationship between the viscosity and the stability of the particles. Thus the present work highlights that by exploiting the intrinsic properties of the solution it is possible to synthesize small and stable nanoparticles without using stabilizer. The Au NPs as synthesized in 20:80 (v/v) glycerol:water mixture were found to be catalytically active and had shown good catalytic activity for reduction of o-nitro aniline.

REFERENCES

1. H. Xiaohua, I. H. El-Sayed, W. Qian and M. A. El-Sayed, "Cancer Cell Imaging and Photothermal Therapy in the Near-Infrared Region by Using Gold Nanorods," Journal of American Chemical Society, Vol. 128, No. 6, 2006, pp. 2115-2120.

2. M. C. Daniel and D. Astruc, "Gold Nanoparticles: Assembly, Supramolecular Chemistry, Quantum-Size-Re-

3. lated Properties, and Applications toward Biology, Catalysis, and Nanotechnology," Chemical Reviews, Vol. 104, No. 1, 2004, pp. 293-346.

4. V. Tjoa, W. Jun, V. Dravid, S. Mhaisalkarad and N. Mathews, "Hybrid Graphene-Metal Nanoparticle Systems: Electronic Properties and Gas Interaction," Journal of Materials Chemistry, Vol. 21, No. 39, 2011, pp. 15593- 15599.

5. L. Dykmana and N. Khlebtsov, "Gold Nanoparticles in Biomedical Applications: Recent Advances and Perspectives," Chemical Society Reviews, Vol. 41, No. 6, 2012, pp. 2256-2282.

6. B. R. Cuenya, "Synthesis and Catalytic Properties of Metal Nanoparticles: Size, Shape, Support, Composition, and Oxidation State Effects," Thin Solid Films, Vol. 518, No. 12, 2010, pp. 3127-3150.

7. L. Rodriguez-Lorenzo, R. Rica, R. A. Alvarez-Puebla, L. M. Liz-Marzan and M. M. Stevens, "Plasmonic Nanosensors with Inverse Sensitivity by Means of Enzymeguided Crystal Growth," Nature Materials, Vol. 11, No. 7, 2012, pp. 604-607.

8. S. Y. Li and M. Wang, "Branched Metal Nanoparticles: A Review on Wet-Chemical Synthesis and Biomedical Applications," Nano Life, Vol. 2, No. 1, 2012, Article ID: 1230002.

9. S. Kapoor, R. Joshi and T. Mukherjee, "Influence of IAnions on the Formation and Stabilization of Copper Nanoparticles," Chemical Physics Letters, Vol. 351, No. 5-6, 2002, pp. 443-448.

10. M. Vaseem, K. M. Lee, D. Y. Kim and Y. -B. Hahn, "Parametric Study of Cost-Effective Synthesis of Crystalline Copper Nanoparticles and their Crystallographic Characterization,"Material Chemistry and Physics, Vol. 125, No. 3, 2011, pp 334-34.

11. K. S. N. Kamarudin and M. F. Mohamad, "Synthesis of Gold (Au) Nanoparticles for Mercury Adsorption," American Journal of Applied Sciences, Vol. 7, No. 6, 2010, pp. 835-839.

12. M. S. Frank, M. C. Nahata and M. D. Hilty, "Glycerol: A Review of Its Pharmacology, Pharmacokinetics, Adverse Reactions, and Clinical Use," Pharmacotherapy, Vol. 1, No. 2, 1981, pp. 147-160.

13. H. W. Felter, "Glycerinum," 1922. http://www.henriettesherbal.com/eclectic/felter/glycerinum.html

14. R. Genc, G. Clergeaud, M. Ortiz and C. K. O'Sullivan, "Green Synthesis of Gold Nanoparticles Using GlycerolIncorporated Nanosized Liposomes," Langmuir, Vol. 27, No. 17, 2011, pp. 10894-10900.

15. K. Singh, K. H. Kate1, V. V. S. Chilukuri and P. K. Khanna, "Glycerol Mediated Low Temperature Synthesis of Nickel Nanoparticles by Solution Reduction Method," Journal of Nanoscience and Nanotechnology, Vol. 11, No. 6, 2011, pp. 5131-5136.

16. E. Nisaratanaporn and K. Wongsuwan, "Preparation of Ultrafine Silver Powder Using Glycerol as Reducing Agent," Journal of Metals, Materials and Minerals, Vol. 18, No. 2, 2008, pp. 1-5.

17. N. Grace and K. Pandian, "One Pot Synthesis of Polymer Protected Gold Nanoparticles and Nanoprisms in Glycerol," Colloids and Surfaces A: Physicochemical and Engineering Aspects, Vol. 290, No. 1-3, 2006, pp. 138-142.

18. V. Sharma, K. Park and M. Srinivasarao, "Colloidal Dispersion of Gold Nanorods: Historical Background, Optical Properties, Seed-Mediated Synthesis, Shape Separation and Self-Assembly," Materials Science and Engineering: R: Reports, Vol. 65, No. 1-3, 2009, pp. 1-38.

19. S. Yang , T. Zhang, L. Zhang, S. Wang, Z. Yang and B. Ding, "Continuous Synthesis of Gold Nanoparticles and Nanoplates with Controlled Size and Shape under UV Irradiation," Colloids and Surfaces A: Physicochemical and Engineering Aspects, Vol. 296, No. 1-3, 2007, pp. 37-44.

20. S. Ghosh, S. Patil, M. Ahire, R. Kitture, D. D. Gurav, A. M. Jabgunde, S. Kale, K. Pardesi, V. Shinde, J. Bellare, D. D. Dhavale and B. A. Chopade, "Gnidia Glauca Flower Extract Mediated Synthesis of Gold Nanoparticles and Evaluation of its Chemocatalytic Potential," International Journal of Nanobiotechnology, Vol. 10, No. 1, 2012, pp. 17-26.

21. L. Chen, C. Wan and Y. Wang, "Chemical Preparation of Pd Nanoparticles in Room Temperature Ethylene Glycol System and Its Application to Electroless Copper Deposition," Journal of Colloid and Interface Science, Vol. 297, No. 1, 297, 2006, pp. 143-150.

22. S. Boopathi, S. Senthilkumar and K. L. Phani, "Facile and One Pot Synthesis of Gold Nanoparticles Using Tetraphenylborate and Polyvinylpyrrolidone for Selective Colorimetric Detection of Mercury Ions in Aqueous Medium," Journal of Analytical Methods in Chemistry, Vol. 2012, No. 1, 2012, Article ID: 348965.

23. G. Oza, S. Pandey, R. Shah, M. Vishwanathan, R. Kesarkar, M. Sharon and M. Sharon, "Tailoring Aspect Ratio of Gold Nano Rods: Impact of Temperature, pH, Silver ions, CTAB Concentration and Centrifugation," Advances in Applied Science Research, Vol. 3, No. 2, 2012, pp. 1027-1038.

24. K. Layek, M. Lakshmi Kantam, M. Shirai, D. NishioHamane, T. Sasaki and H. Maheswaran, "Gold Nanoparticles Stabilized on Nanocrystalline Magnesium Oxide as an Active Catalyst for Reduction of Nitroarenes in Aqueous Medium at Room Temperature," Green Chemistry, Vol. 14, No. 11, 2012, pp. 3164-3174.

25. A. Leelavathi, T. U Bhaskara Rao and T. Pradeep, “Supported Quantum Clusters of Silver as Enhanced Catalysts for Reduction,” Nanoscale Research Letters, Vol. 6, No. 2, 2011, pp. 123-132.

26. N. Pradhan, A. Pal and T. Pal, “Catalytic Reduction of Aromatic Nitro Compounds by Coinage Metal Nanoparticles,” Langmuir, Vol. 17, No. 5, 2001, pp 1800-1802.

Chapter 7

Thermal Conductivity Enhancement of Epoxy by Hybrid Particulate Fillers of Graphite and Silicon Carbide

Krishnamachar Srinivas[1*], Mysore Siddalingappa Bhagyashekar[2]

[1]Department of Mechanical Engineering, Don Bosco Institute of Technology, Bangalore, India

[2]Department of Mechanical Engineering, Raja Rajeshwari College of Engineering, Bangalore, India

ABSTRACT

This paper reports thermal conductivity studies carried out on room temperature cure (RT) epoxy resin (LY556 + HY951) containing three different particulate fillers such as Graphite (Gr) a soft material, Silicon carbide (SiC) a hard material and a hybrid graphite & silicon carbide (Gr-SiC). The weight fractions of three different fillers were varied from 10 wt% to 40 wt% in steps of 10%. Increased filler fraction increases the thermal conductivity of epoxy composites for all the three types of fillers. The results show that the synergic effect of hybrid filler (Gr-SiC) improves the thermal conductivity of epoxy composites compared to that of Graphite or Silicon carbide. The improvement in thermal conductivity for the epoxy hybrid composite containing 20% SiC, 20% Gr and 60% epoxy is 136% when compared with neat epoxy. Significant improvement in the thermal conductivity is observed for 40% filled epoxies. The experimental results were compared with analytical models such as Maxwell, Hashin, Hamilton-Crosser, Nielsen, and Cheng-Vachon. The predicted thermal conductivity by analytical models is in agreement with experimental results.

KEYWORDS

Hybrid Composites, Particulate Epoxy Composite, Thermal Conductivity

1. INTRODUCTION

Polymers in general have low thermal conductivity due to low atomic density, chemical bonding, complex crystal structure [1] and anharmonicity in molecular vibrations [2] . The heat transfer in non metals occurs by phonon or lattice vibrations causing thermal resistance [3] . Epoxy as a matrix has a wide applications in the field of engineering. They have some of the favorable properties such as compatibility with many materials, higher thermal and chemical resistance cheaper than other thermosets. However, due to its cross linked structure, epoxies have disadvantages like brittleness, low impact strength, low wear resistance [4] . Epoxies are poor conductors of heat and current. Hence to improve their thermal conductivity, one of the routes is to reinforce particulate fillers. Excellent conductive properties of epoxies have applications in miniature devices managing heat transfer. To improve thermal conductivity, the thermal resistance has to be minimized. A good interfacial adhesion by modifying the filler surface reduces thermal resistance [5] . Further, when the thermal conductivity of the filler is more than 100 times the thermal conductivity of the polymer matrix, there will not be substantial increase in thermal conductivity of the composite [3] . To overcome this, the aspect ratio of the filler has to be increased so as to create conductive paths [6] [7] . The conductive paths can be created by increasing the volume fraction of fillers, increasing the size of particles and use of hybrid fillers. The fillers with large aspect ratio are capable of creating a conductive network [8] [9] . Hence, the present study looks into the modification of epoxy, by incorporation of particulates of Gr, SiC and hybrid (Gr-SiC) at various weight fractions to improve thermal conductivity. The Gr is considered as one of the best conductive filler due to its ability to disperse in a matrix, low cost and good thermal conductivity [1] . SiC microfiller is a choice material for high temperature and high power applications, as it has high thermal conductivity and low thermal expansion coefficient [10].

2. EXPERIMENTAL DETAILS

2.1. Materials Selecting a Template

Micro particulates of Graphite (Gr) supplied by SD fine chem ltd, India having size less than 60 μm and particulates of Silicon carbide (SiC) supplied by Gran Silica India, having size less than 20 μm (Figure 1) were used as fillers. The host matrix is an epoxy resin of type LY556 (RT cure) and an amine hardener HY951 supplied by Huntsman India Ltd. The resin and hardener were mixed in the weight ratio 10:1.

2.2. Specimen Preparation and Characterization

A metallic mould was fabricated to give specimens of 2 mm thick and 25 mm diameter which is coin shaped as shown in Figure 2. Before casting, the mould was cleaned thoroughly and a gel coat was applied to the mould. The particulates were preheated to remove any moisture present in it. The predetermined amounts of fillers were dispersed in epoxy resin, stirring continuously till the time of pouring. The cast specimens were allowed to cure in the mould for 24 hours at room temperature. The room temperature cured specimens were post cured for the following schedule of 50°C for 30 minutes, 70°C for 60 minutes and 85°C for 120 minutes. To begin with, unfilled epoxy was cast first. Further, Gr-epoxy, SiC-epoxy and hybrid (Gr-SiC) epoxy composites were cast. The weight fractions of particulates of Gr and SiC were varied from 10% to 40% (maximum) in steps of 10%. In the next step, hybrid (Gr-SiC) filled epoxy composites were cast containing the filler weight fraction as given below.

Gr fraction kept at 5%, SiC fraction varied from 5% to 35% in steps of 10%.

Gr fraction kept at 10%, SiC fraction varied from 10% to 30% in steps of 10%.

SiC fraction kept at 5%, Gr fraction varied from 5% to 35% in steps of 10%.

SiC fraction kept at 10%, Gr fraction varied from 10% to 30% in steps of 10%.

Both SiC and Gr fraction varied equally up to 40% in steps of 10%.

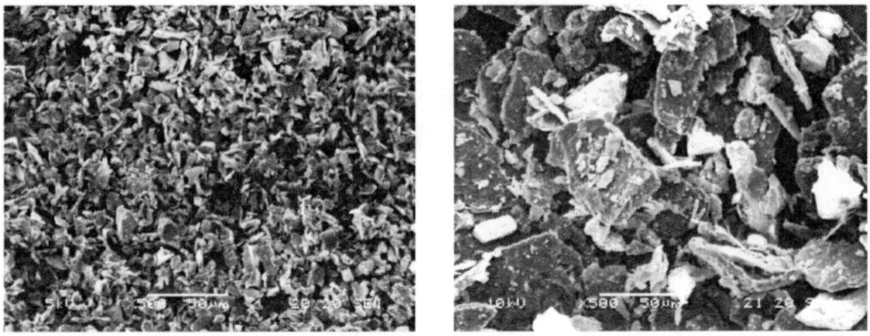

(a) (b)

Figure 1. SEM Photographs of particulates of Gr and SiC. (a) SiC particles; (b) Gr particles.

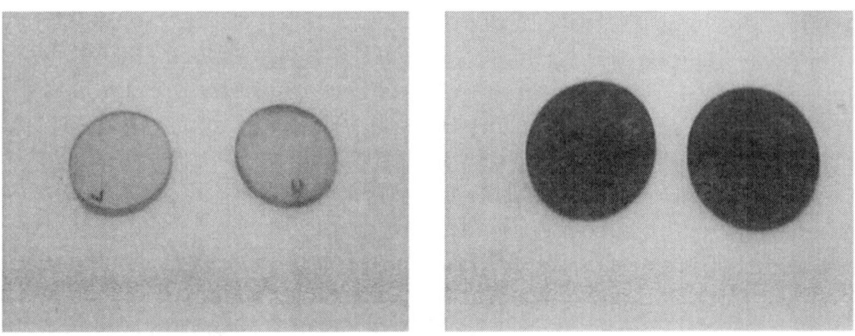

(a) (b)

Figure 2. Cast specimens of epoxy and (Gr-SiC) epoxy hybrid composite. (a) Unfilled epoxy; (b) (Gr-SiC) Hybrid epoxy composite.

The different types of cast composites along with their codes are presented in Table 1. For ex composite 5S35G-Ep represents particulate content of 5% SiC, 35% Gr and 60% epoxy. Similarly a composite having code 10G30S-Ep has 10% Gr, 30% SiC and 60% epoxy.

2.3. Measurement of Thermal Conductivity

Thermal conductivity measurements were carried out by an instrument developed by Indian national academy of science. The schematic diagram of the set up is shown in Figure 3. The thermal conductivity measurements were based on steady state method. One directional heat flow was considered for measurement of thermal conductivity.

2.4. Description of the Apparatus

A heater made of six 120-Ohm 1/4 W resistors in parallel is placed in the copper cup of 25 mm diameter. It is connected to a constant current source. Two cast specimens are clamped between the copper cup and two square copper plates. Two junctions of a copper constantan thermocouple are fixed to upper half of the copper heater cup and to the upper square plate. Another pair of junctions of the thermocouple is fixed to bottom edge of the bottom half of the heater cup and bottom square plate. The leads from the thermocouple are connected to the differential amplifier. The specimens are fixed in between the square plate and the copper cup. The set up is connected to a constant current source. The outputs from the thermocouples are connected to a DC differential amplifier. Initially the current is set at 140 mA. The steady state is obtained after 45 mins. The output voltage is measured on a digital multimeter for thermocouple 1 and the process is repeated for thermocouple 2. Further the current is increased to 160 mA and the measurements are repeated. The measurements are repeated for five different current settings and two specimens were used for each experiment.

The experimental thermal conductivity is evaluated using the equation

$$Q = -KA\left[\frac{\Delta T}{d}\right] \tag{1}$$

where, K is the thermal conductivity of composite in W/mK;

Q is the heat supplied in watts;

A is the area of the specimen in m^2;

d is the thickness of the specimen in m.

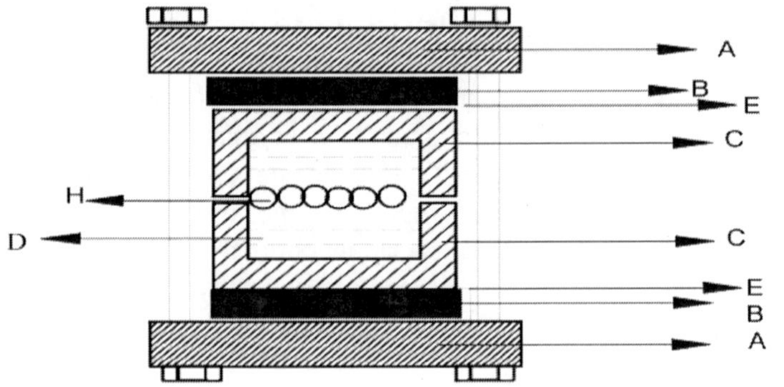

Figure 3. Schematic Diagram of the set up. A: Copper base plate each 50 mm square 5 mm thick; B: Specimen 25 mm dia and 2 mm thick; C: Two halves of a copper cup 25 mm dia, outer height 5 mm and inner height 3 mm; D: Aluminium foils pads; H: Heater made up of six 120 ohm 1/4 W resistors in parallel; E: Cu-Constantan thermocouples.

2.5. Theoretical Models

The following analytical models were used to predict the thermal conductivity of epoxy composites [11] [12] . The models considered were Maxwell equation, cylinder assemblage model developed by Hashin Nielsen's equation and series equation, which is an inverse rule of mixtures which gives lower bound values of thermal conductivity for particulate filled composites [12] .

a) Maxwell Equation

$$\frac{K}{Kc} = 1 + \left[\frac{3(Kd - Kc)}{(Kd + 2Kc)} \right] \Phi \tag{2}$$

where K, Kc and Kd are the thermal conductivities of composites, matrix and fillers Φ is the volume fraction of filler.

b) Cylinder assemblage model: Cylinder Assemblage model was developed by Hashin

$$K_T = K_{TM} \left\{ K_{TM} V_M + K_{TF} \left(1 + V_F\right) / K_{TM} \left(1 + V_F\right) + K_{TF} V_M \right\} \tag{3}$$

Table 1. Thermal conductivity of epoxy and its composites.

	Code	%Ep	%SiC	%Gr	Expt (W/mK)	Series (W/mK)	Maxwell (W/mK)	Hashin (W/mK)	Neilson (W/mK)
Epoxy and 10% Filled epoxy	Ep	100	0	0	0.30	0.30	0.30	0.30	0.30
	10G-Ep	90	0	10	0.32	0.33	0.39	0.37	0.39
	10S-Ep	90	10	0	0.33	0.33	0.39	0.37	0.38
	5S5G-Ep	90	5	5	0.33	0.39	0.37	0.38	0.33
20% Filled epoxy	20G-Ep	80	0	20	0.46	0.37	0.48	0.47	0.51
	20S-Ep	80	20	0	0.41	0.37	0.49	0.45	0.49
	5S15G-Ep	80	5	15	0.46	0.48	0.45	0.52	0.40
	5G15S-Ep	80	15	5	0.46	0.48	0.45	0.52	0.40
	10S10G-Ep	80	10	10	0.51	0.48	0.45	0.50	0.37
30% Filled epoxy	30G-Ep	70	0	30	0.54	0.43	0.57	0.59	0.7
	30S-Ep	70	30	0	0.51	0.43	0.58	0.56	0.63
	5S25G-Ep	70	5	25	0.52	0.57	0.56	0.67	0.43
	5G25S-Ep	70	20	5	0.52	0.57	0.56	0.67	0.43
	10S20G-Ep	70	10	20	0.56	0.54	0.55	0.72	0.45
	10G20S-Ep	70	20	10	0.54	0.54	0.55	0.64	0.45
	15G15S-Ep	70	15	15	0.58	0.60	0.59	0.69	0.45
40% Filled epoxy	40G-Ep	60	0	40	0.64	0.50	0.66	0.73	1.04
	40S-Ep	60	40	0	0.68	0.50	0.70	0.70	0.82
	5S35G-Ep	60	5	35	0.64	0.66	0.70	0.86	0.50
	5G35S-Ep	60	35	5	0.62	0.66	0.70	0.86	0.50
	10S30G-Ep	60	10	30	0.62	0.66	0.7	1.02	0.53
	10G30S-Ep	60	30	10	0.61	0.66	0.70	0.84	0.53
	20G20S-Ep	60	20	20	0.71	0.71	0.70	0.90	0.50

where, K_T, K_{TM} and K_{TF} are thermal conductivities of composite, matrix and filler.

V_F, V_M are the volume fraction of filler and matrix.

c) Nielsen's Equation

$$\frac{K_c}{K_m} = \frac{1 + AB\varphi}{1 - \phi B\varphi}$$

Where

$$B = \frac{k_f / k_m - 1}{k_f / k_m + A} \qquad (4)$$

where A is the geometry of the particles = 1.5 and ϕ_{max} (maximum packing fraction)

d) Inverse rule of mixtures/Series

$$\frac{1}{K_T} = \frac{V_f}{K_f} + \frac{V_m}{K_m} \qquad (5)$$

The above equation is derived from rule of mixtures and does not consider voids.

3. RESULTS AND DISCUSSION

3.1. Effect of Gr/SiC on Thermal Conductivity of Gr-Epoxy/SiC-Epoxy Composites

The experimental thermal conductivity and predicted by analytical models of all types of epoxy composites is listed in Table 1. Figure 4(a) and Figure 4(b) presents the thermal conductivity of epoxy composites filled with particulates of Gr and SiC. Neat epoxy exhibits a thermal conductivity of 0.30 W/mK. For filled Gr/SiC epoxy, up to 10% filler fraction, the rise in thermal conductivity is not significant. The epoxy composites containing 10% fillers, 10G-Ep and 10S-Ep exhibits a thermal conductivity of 0.32 W/mK and 0.33 W/mK respectively which is an insignificant improvement over neat epoxy. This may be due to low filler fraction and hence effective conductive paths are not created in the composite. Further addition of particulates of graphite increases thermal conductivity of graphite epoxy composites linearly. The thermal conductivity increases to 0.46 W/mK and 0.54 W/mK for 20G-Ep and 30G-Ep respectively. Maxima of 0.64 W/mK was observed for 40G-Ep which is an improvement of 113% over unfilled epoxy. The effect of adding SiC to epoxy is shown inFigure 4(b). The SiC-epoxy composites have trends similar to Gr-epoxy composites. However, 20S-Ep and 30S-Ep have thermal conductivity of 0.41 W/mK and 0.51 W/mK respectively which is marginally lower than Gr-epoxy under similar loading. However 40S-Ep, exhibits a thermal conductivity of 0.68 W/mK. This suggests that even though the thermal conductivity of SiC (280 W/mK) is higher than Gr (85 W/mK), SiC-Ep composites have lower thermal conductivity than Gr-Ep composites. The factors such as larger particle size of Gr (20 - 60 μm) (Figure 1(b)) irregularity in shape, easy dispersibility [11] favors enhancement of thermal conductivity, where as SiC particulates are smaller in size (6 - 12 μm) uniform (Figure 1), are not beneficial in creating conductive paths [8] [9] .

Comparing thermal conductivity experimental with thermal conductivity predicted by analytical equations, it is seen that, series equation predicts lower thermal conductivity for all filler loadings where as Hashin and Maxwell equations are closer to experimental values. The prediction by Nielsen's equation is in agreement with experimental value up to 20% filler loading and deviates after 30% filler fraction. This is in agreement with [3] [5] , where the Nielsen's estimation at lower volume fraction matches the experimental values and tends to over estimate the thermal conductivity of composites at higher volume fraction.

3.2. Effect of Gr on Thermal Conductivity of Hybrid (Gr-SiC)-Epoxy Composites

The effect of varying Gr on hybrid (Gr-SiC) epoxy composites is shown in Figure 5(a) and Figure 5(b). Figure 5(a) shows the thermal conductivity of epoxy composites, where SiC was unvaried at 5% for all composites, Gr varied from 5% to 35% in steps of 10%. Increased Gr fraction increases the thermal conductivity of hybrid (Gr-SiC) epoxy composites. Thermal conductivity increases to 0.46 W/mK for 5S15G-Ep, 0.52 W/mK for 5S25G-Ep and 0.64 W/mK for 5S35G-Ep composite.

The effect of Gr where SiC is held at 10% and Gr is varied from 10% to 30% is shown in Figure 5(b). These composites present a trend similar to previous compositions. However, these composites exhibits marginally lower conductivity values than composites having 5% SiC unvaried composites. Maximum thermal conductivity of 0.62 W/mK was observed for 10S30G-Ep. The SEM of the surfaces of 5S35G-Ep and 10S30G-Ep is shown in Figure 6(a) and Figure 6(b) respectively. The few bright spots seen in the images are particulates of SiC.

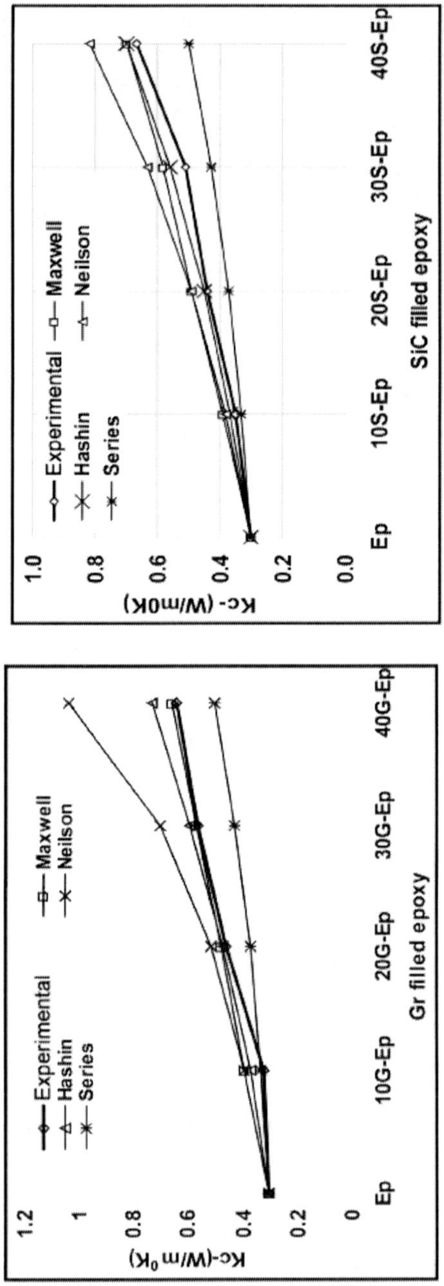

Figure 4. Thermal conductivity of Gr/SiC filled epoxy. (a) Gr filled epoxy; (b) SiC filled epoxy.

3.3. Effect of SiC on Thermal Conductivity of Hybrid (Gr-SiC)-Epoxy Composites

Figure 7(a) and Figure 7(b) illustrates the effect of varying SiC and holding Gr constant at 5%, 10% respectively on hybrid (Gr-SiC) epoxy composites. Thermal conductivity increases linearly for all type of composites. The thermal conductivity increases to 0.52 W/mK for 5G25S-Ep and 0.62 W/mK for 5G35S-Ep. Similarly 10G20S-Ep and 10G30S-Ep exhibits thermal conductivity of 0.54 W/mK and 0.62 W/mK respectively. The SEM images of 5G35S-Ep and 10G30S-Ep composites are shown in Figure 8(a) and Figure 8(b). As SiC fraction is more for these composites, more bright spots are seen, indicating well distribution of SiC.

3.4. Effect of Equal Fraction of Gr and SiC on Thermal Conductivity of Hybrid (Gr-SiC)-Epoxy Composites

The effect of hybrid (Gr-SiC) fillers on epoxy where both fillers are equally varied is shown in Figure 9. The thermal conductivity of 10G10S-Ep is found to be 0.51 W/mK which is highest among 20% filled composites. Similarly 15G15S-Ep exhibits a thermal conductivity of 0.58 W/mK, highest among 30% filled epoxy. Similarly the thermal conductivity of 20G20S-Ep is 0.71 W/mK, highest among 40% filled composites. Further, it is also observed that for the same filler loadings, the thermal conductivity varies. As filler loading increases, number of interfaces increases and thermal conductivity in a polymer composite is dependent on good interfacial adhesion. For a poor interfacial adhesion phonon scattering increases, and decreases the thermal conductivity. Further different particle sizes vary the interfaces along the heat transfer path [13] . From the above it is evident that synergy exists between hybrid (Gr-SiC) when both the fillers are loaded equally.

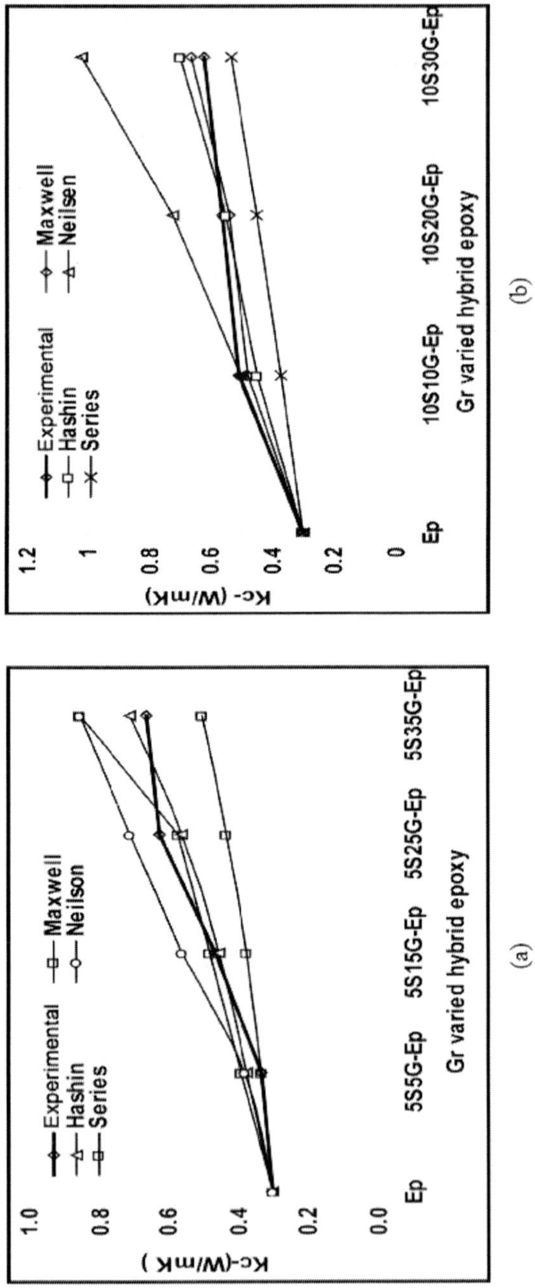

Figure 5. Thermal conductivity of graphite rich hybrid (Gr-SiC) epoxy composites. (a) SiC constant at 5%, Gr varied from 5% - 35%; (b) SiC constant at 10%, Gr varied from 10% - 30%.

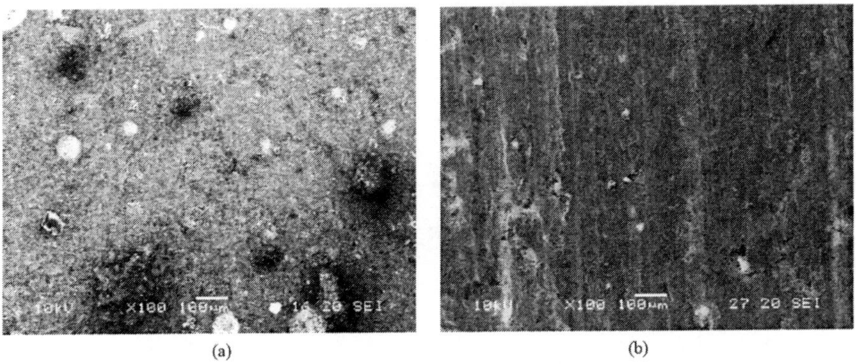

Figure 6. SEM images of surface of graphite rich hybrid (Gr-SiC) epoxy composites. (a) 5S35G-Ep; (b) 10S30G-Ep.

4. CONCLUSION

Thermal conductivity increases linearly with increased filler fraction for all composite systems irrespective of fillers. Thermal conductivity show significant increase after 20% filler fraction. Comparing Gr-Ep and SiC-Ep composites, for the same filler fraction, Gr-Ep composites have higher thermal conductivity than SiC-Ep composites except for 40S-Ep which has higher thermal conductivity than 40G-Ep. Hybrid composites having equal proportion of filler fraction i.e. 10G10S-Ep, 15G15S-Ep and 20G20S-Ep have thermal conductivity highest in 20%, 30% and 40% filled epoxy composites. The hybrid composite 20G20S-Ep has a thermal conductivity of 0.71 W/mK which is highest among all the composite considered in this study and is an improvement of 136% over a neat epoxy. The experimental results were compared with the analytical models. The experimental results exhibit good correlation with Maxwell's equation.

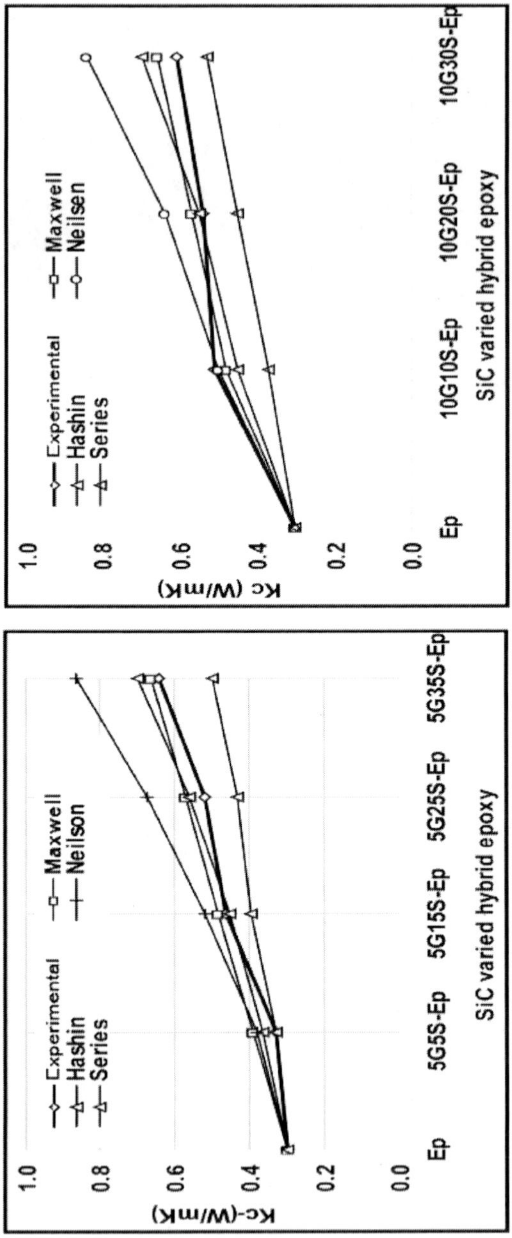

Figure 7. Thermal conductivity of SiC rich hybrid (Gr-SiC) epoxy composites. (a) Gr constant at 5%, SiC varied from 5% - 35%; (b) Gr constant at 10%, SiC varied from 10% - 30%.

Figure 8. SEM images of surface of graphite rich hybrid (Gr-SiC) epoxy composites. (a) 5G35S-Ep; (b) 10G30S-Ep.

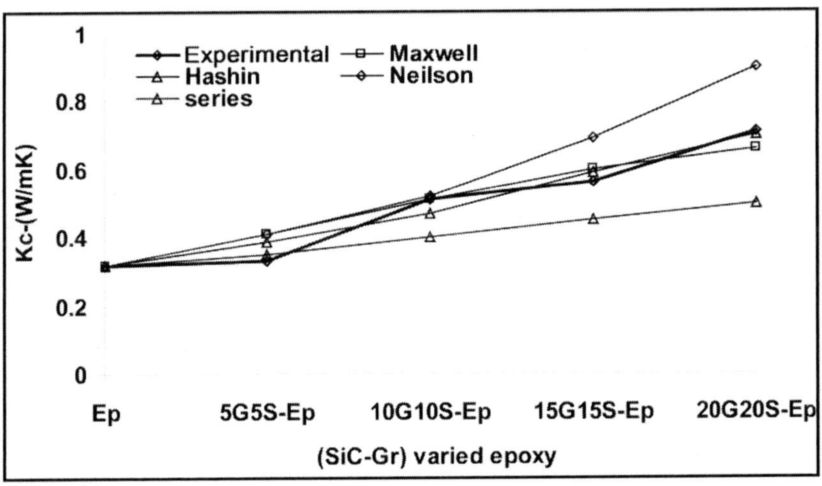

Figure 9. Effect of fillers on thermal conductivity of epoxy composites. Both Gr and SiC varied equally.

ACKNOWLEDGEMENTS

Krishnamachar Srinivas, Mysore Siddalingappa Bhagyashekar The authors thank the management of Don Bosco Institute of Technology, RajaRajeswari College of Engineering Bangalore and Indian Academy Of Science. Bangalore.

REFERENCES

1. Ebadi-Dehaghani, H. and Branch, M.N.S. (2012) Thermal Conductivity of Nanoparticles Filled Polymers Islamic Azad University Iran Smart Nanoparticles Technology. 519-540.

2. Tsekmes, I.A., Kochetov, R., Morshuis, P.H.F. and Smit, J.J. (2013) Thermal Conductivity of Polymeric Composites: A Review. 2013 IEEE International Conference on Solid Dielectrics, Bologna, 30 June-4 July 2013, 678-681.

3. Leea, G.-W., Parka, M., Kima, J., Leeb, J.I. and Yoonb, H.G. (2006) Enhanced Thermal Conductivity of Polymer Composites Filled with Hybrid Filler. Composites: Part A, 37, 727-734.

4. Wang, M.C., Zhang, Z.G. and Sun, Z.J. (2009) The Hybrid Model and Mechanical Properties of Hybrid Composites Reinforced with Different Diameter Fibers. Journal of Reinforced Plastics and Composites, 28, 257.

5. Kanga, S., Honga, S.I., Choeb, C.R., Parkb, M., Rimb, S. and Kimb, J. (2001) Preparation and Characterization of Epoxy Composites Filled with Functionalized Nanosilica Particles Obtained via Solgel Process. Polymer, 42, 879-887.

6. Luyt, A.S., Molefi, J.A. and Krump, H. (2005) Thermal Mechanical and Electrical Properties of Copper Filled Low Density Ald Linear Low Density Polyethylene Composites. Polymer Degradation and Stability, 1-8.

7. Yung, K.C., Wang, J. and Yue, T.M. (2008) Thermal Management for Boron Nitride Filled Metal Core Printed Circuit Board. Journal of Composite Materials, 42, 2615.

8. Zhou, W.Y., Wang, C.F., An, Q.L. and Ou, H.Y. (2008) On Thermal Properties of Heat Conductive Silicone Rubber Filled with Hybrid Fillers. Journal of Composite Materials, 42, 173.

9. Mamunya, Ye.P., Davydenko, V.V., Piss, P. and Lebedev, E.V. (2002) Electrical and Thermal Conductivity of Polymers Filled with Metal Powders. European Polymer Journal, 38, 1887-1897.

10. Zhou1, T., Wang, X., Cheng, P., Wang, T., Xiong, D. and Wang, X. (2013) Improving the Thermal Conductivity of Epoxy Resin by the Addition of a Mixture of Graphite Nanoplatelets and Silicon Carbide Microparticles. Express Polymer Letters, 7, 585-594.

11. Reine, B., Tomaso, J.D., Dusserre, G. and Olivier, P.A. (2012) Study of Thermal Behaviour of Thermoset Polymer Matrix Filled with Micro and Nano Particles. Proceedings of ECCM15, Italy, 24-28 June 2012.

12. Pal, R. (2007) New Models for Thermal Conductivity of Particulate Composites. Journal of Reinforced Plastics and Composites, 26, 643.

13. Hong, J.-P., Yoon, S.-W., Hwang, T.-S., Lee, Y.-K., Won, S.-H. and Nam, J.-D. (2010) Inter Phase Control of Boron Nitride/Epoxy Composites for High Thermal Conductivity. Korea-Australia Rheology Journal, 22, 259-264.

Chapter 8

Thermal Stability and Hot Carrier Dynamics of Gold Nanoparticles of Different Shapes

Yasser A. Attia[1,2,3], Tariq A. Altalhi[2], Adil A. Gobouri[2]

[1]National Institute of Laser Enhanced Sciences, Cairo University, Giza, Egypt

[2]Department of Chemistry, Faculty of Science, Taif University, Taif, Kingdom of Saudi Arabia

[3]Department of Physical Chemistry, Faculty of Chemistry, Santiago de Compostela University, Santiago de Compostela, Spain

ABSTRACT

Anisotropic shapes of gold nanoparticles are prepared using a modified seed method in the presence of silver ions or clusters in order to study the thermal stability and the dynamics of the hot carriers induced by femtosecond laser pulses. Although gold nanospheres are stable towards thermal treatment, the decomposition of the gold nanorods into spherical nanoparticle aggregates upon thermal treatment is mechanistically different from the case of nanoprisms. Great enhancement of thermal stability is achieved by modifying the surface of the nanoparticles by adding specific amounts of polyvinyl pyrrolidone (PVP) after preparation of gold particles of different shapes capped with cetyltrimethylammonium bromide (CTAB). The surface plasmon resonance spectra of the gold nanostructures are used to monitor their morphological changes. In regards to the hot carrier dynamics, it is found that the phonon-phonon (ph-ph) coupling is much slower in dots than in rods and prisms while electron-phonon (e-ph) coupling is almost the same in these particles.

KEYWORDS

Thermal Reshaping, Stability, Relaxation Dynamics, Anisotropic Gold nanoparticles, Femtosecond laser, capping Materials

1. INTRODUCTION

One of the desired goals is to overcome the challenge of controlling the morphology of the nanoparticles, and to be able to fabricate and stabilize them in the desired form. The electrical, thermodynamic, and physicochemical properties of metal nanoparticles depend sensitively on particle size and shape [1] [2] . There are different methods to prepare different shapes of gold nanoparticles but the most common one is the seed-mediated method which is developed by Murphy's group and further improved by El-Sayed's group [3] -[9] .

Controlling the shape, morphology, and size can be useful in understanding of the electron dynamics and the relaxation processes in gold nanoparticles, which is required because of their fast optical response. This is very important for many potential applications such as optical switching, electronic devices, and biomedical sensing and diagnostics applications [8] -[14] . Moreover, better understanding of properties such as electrical and thermal conductivity and superconductivity of metal nanoparticles could be achieved by studying the relaxation dynamics.

In gold nanoparticles, because of the large difference in the heat capacity of the electrons and the phonons, they can be treated as two coupled subsystems. Before excitation, the electrons occupy the energy states below the Fermi level. Upon excitation of metallic thin films, or gold nanoparticles by ultrafast laser pulses, the energy is transferred to the electrons by absorption of photons via interband and intraband transition. The relaxation processes start to thermalize the electrons via electron-electron scattering and electron scattering with the surface. This thermalization process occurs in a few hundreds of femtoseconds [14] -[18] . The second step in the relaxation process is the transfer of the energy to the lattice via electron-phonon interaction. The last step in the relaxation is the energy transfer to the surrounding medium. The electron?electron and electron-phonon coupling processes and their dependence on the particle size were studied extensively [19] -[28] , but how it affects the particle shape are not completely understood which is a target in this work.

Although there are several studies investigating the thermal stability of the gold nanoparticles in a solution, they are scattered and do not include the effect of all the shapes and surface coatings [29] -[33] . Some examples are: El-Sayed's group studied thermal stability of electrochemically prepared gold nanorods encapsulated inside the mixture of cetyltrimethylammonium bromide (CTAB) and tetraoctylammonium bromide (TDAB) rod shape micelles [21] and the mean aspect ratio of these nanorods in aqueous solution decreases with increasing the temperature. In another study, Petrova et al. [29] observed that gold nanorods at the highest temperature examined (250˚C) would transform into spheres within one hour. Al-Sherbini et al. [31] [32] studied the effect of the thermal heating and UV-light on gold nanorods with different aspect ratios in water/glycerol mixtures.

The objectives of this study are twofold that are continuation to our previous results [34] -[36] ; 1) to investigate the role of shape and capping material on the thermal decomposition of gold nanoparticles as the stability enhancement is of interest for practical applications and 2) to investigate the dynamics of the hot carriers in the gold nanoparticles of different shapes, induced by femtosecond laser pulses.

2. EXPERIMENTAL SECTION

For the preparation of different shapes of gold nanoparticles, seeds and growth solutions were made as described below [34] -[36] :

Seeds solution: 2.5 ml of 0.2 M CTAB solution was mixed with 2.5 ml of 0.5 mM $HAuCl_4$. To the stirred solution, 30 μl of ice-cold 0.1M $NaBH_4$ was added, which resulted in the formation of a brownish yellow solution. Vigorous stirring of the seeds solution was continued for 2 minutes.

Growth solution: Different amounts (0, 0.1 and 0.25 ml) of 4mM $AgNO_3$ solution were added to 2.5 ml of 0.2 M CTAB solution. Then, 2.5 ml of 1 mM $HAuCl_4$ was added and after gentle mixing of the solution, 35 µl of 0.1 M ascorbic acid was added. Ascorbic acid as a reducing agent changes the growth solution from dark yellow to colorless within 10 minutes.

Next, the seeds solution was added to the growth solution. The color of the solution gradually changed within 10 - 20 minutes. To study thermal stability of gold nanoparticles, the change in the absorption spectrum of the gold nanoparticles solution was determined at different temperatures ranging from 40˚C to 220˚C in water bath (Transonic 46 O/H, USA). After the addition of a few drops of glycerol, the sample was heated above 100˚C in an oil bath in order to determine the decomposition temperature of the most stable micelles capping the gold nanoparticles. The same step was repeated by mixing nanoparticles with a solution of PVP. Thermal stability of gold nanoparticles has been followed by measuring the changes in absorption spectra as a function of time using the Perkin-Elmer Lambda 40 spectrometer. The particles were characterized by Transmission Electron Microscopy (TEM) with a Philips CM20 microscope, operating at an accelerating voltage of 200 KV.

2.1 Femtosecond time-Resolved transient absorption Technique

The amplified Ti:sapphire laser system (spectraphysics) consists of two main components: One part is a passively mode-locked Ti:sapphire oscillator which is pumped by a diode-pumped, frequency-doubled Nd:Vandate laser (Coherent Verdi). The oscillator produces femtosecond laser pulses with pulse energies in the range of nJ. The output wavelength of the oscillator is set to 800 nm. The second main component is a regenerative amplifier which is pumped by a diode-pumped, frequency-doubled Nd:Vandate laser (Coherent Verdi) at repetition rate of 1 kHz. Since ultrashort laser pulses have high peak powers and can destroy the lasing medium the pulses generated in the oscillator part have to be stretched first before they can be injected into the amplifier. This is done by chirping the pulse using a grating. Later, after amplification another grating arranged in the opposite way compresses the pulse again to its original duration in the pulse compressor. This technique of amplifying the pulse energy is called chirped pulse amplification (CPA). The timing of the pulse injection (seeding) into the amplifier and ejection of the amplified pulse into the pulse compressor is controlled by a Pockels cell. The final outputs of this laser system are laser pulses of about 100 fs duration (full width at half maximum (FWHM)), with energy of 0.9 mJ at 800 nm at a repetition rate of 1 kHz.

A schematic overview of the optics of the femtosecond transient absorption spectrometer is given in Scheme 1. The output pulse from the amplified Ti:apphire laser system is split into three parts. A small part (4%) generates a white light continuum in a 1 mm sapphire plate which was used as the probe light. The pump source can be 400 and 266 nm laser pulses by SHG and third harmonic generation (THG). The pump pulse is delayed with respect to the probe pulse with an optical delay. Moving the delay line forward will decrease the distance the pump pulse has to travel and causes the pulse to arrive earlier in time with respect to the probe pulse whose travel distance remains constant. The delay line is computer controlled and allows a time resolution of 21 fs. The energies of the pump and probe pulses are adjusted with neutral density filters.

The probe light splits into a signal and a reference beam and the signal beam is overlapped at the sample position. The signal and reference beams are then focused into fiber optic cables and coupled into a monochromator for single wavelength measurements or a spectrograph for spectral measurements. The reference beam is split off beforehand and is directed towards a separate fiber optics cable.

The samples consisting of mainly transparent solutions are placed in a rectangular stationary cell or in cylindrical cell that can be rotated with a speed of 1000 rpm in order to ensure that a different part of the solution is exposed with each laser shot. Alternatively the solutions can be flowed through a rectangular cell by a micro- pump at a speed of roughly 125 ml/s. Experiments are performed at room temperature. All cells have an optical path length of 2 mm as the overlap of the probe and pump beam can only be achieved over this distance.

For recording single wavelength kinetic traces, the excitation beam is modulated by an optical chopper (HMS 221) with half the laser frequency (500 Hz). Hence only every other probe pulse monitors the changes in the sample induced by the pump pulse. This then causes an intensity modulation of the probe pulses, which varies with the same frequency of 500 Hz. A lock-in-amplifier can then be used to detect and measure very small Ac signals down to a few nanovolts. Accurate measurements may even be made when the signal is obscured by noise sources thousands of times larger like fluctuations in the laser light intensity of pump and probe pulses. The lock-in-amplifier uses a technique known as phase-sensitive detection to single out the component of the signal at the specific reference frequency with which the pump pulse are modulated by the optical chopper. Noise signals at frequencies other than the reference frequency are then rejected and do not affect the measurement.

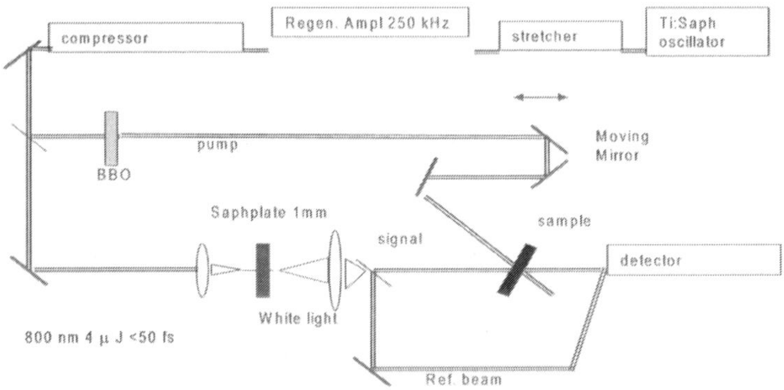

Scheme 1. Description of pump-probe transient setup.

3. RESULTS AND DISCUSSION

3.1. Shape and Morphology of Gold Nanoparticles

The seed method was successfully modified here in order to get more precise control in the shape of the gold nanoparticles. Anisotropic shapes of gold nanoparticles in Figure 1 and Figure 2 were formed by using different concentrations of silver ions and seeds solutions. The average size of the particles is determined from the transmission electron microscopic (TEM) and scanning tunneling microscope (STM) images and is found to be 5.0 ± 1.6 nm for the spheres, and rod length is 40.87 ± 6.1 nm, and its width 12.0 nm with aspect ratio a.r. = 3.4, and the prism is 41.12 ± 5.49 nm, and the thickness is about 9.1 nm (see figure 2).

Our results indicate that the ratio of the Ag^+ concentration to the seed concentration is one of the key parameters for controlling the shape of the particles, e.g. spheres, rods, prisms or snapped prisms. Recently a new proposed mechanism for the formation of anisotropic gold nanoparticles was published that takes into account the reduction of Ag ions by ascorbic acid forming intermediate small Ag clusters, which can then act as a catalyst for the formation of gold nanoparticles by their adsorption onto the surface of the gold seeds [34] [36] -[38] . By controlling the concentration of silver ions that reduced to silver clusters and adjust the ratio with seeds concentration, gold spheres are formed as this ratio equal zero, rods as the ratio equal 1 and prisms are formed as the ratio equal 3 [34] . After controlling the shapes of gold nanostructures, the thermal stability of the formed particles was studied.

3.2. Thermal Stability of Gold Nanoparticles

The prepared gold nanoparticles are capped with only one surfactant (CTAB), and the anisotropic shape resulted by adding Ag^+ ions or clusters during particle growth [34] . It is important to understand the effect of surfactant structure on the stability of the formed micelles which represent the capping material of the gold particles. It is concluded that the Kraft temperature of the micelles depends on their chemical composition and the nature of the capping materials [28] .

Figures 3(a)-(d) show the thermal stability of gold nanoparticles with different shapes. Gold nanospheres prepared in the absence of any Ag-ions or clusters, are not showing any change in the maximum absorption (figure 3(a)) under thermal heating up to 180°C. That is because the spherical shape of gold particles has the most stable facets {111} and these spheres capped with stable micelleses of CTAB.

However, gold nanorods (figure 3(b)) and nanoprisms (figure 3(c)) here are decomposed under the thermal effect ranging from 40°C to 180°C. Since the gold particles are soluble in water which boil at 100°C, few drops of glycerol has been added, and the sample was heated to 180°C in an oil bath in order to determine the decomposition temperature of the most stable micelles capping the gold nanoparticles (Figure 4). Normalized absorption spectra (figure 3(d)) show the difference in stability of the different shapes.

At each temperature, the change in the absorption spectrum was followed as a function of time. The rate of the change in the shape at each of these temperatures from that present at room temperature was determined by following the rate of the absorbance change of the absorption maximum of longitudinal plasmon band SP_L. From the values of the rate constants (first-order decay) at different temperatures, the activation energy (E_{act}) of the shape transformation can be calculated [28] . The activation energy for gold nanorods with aspect ratio 3.4 (7.41 ± 0.1 kJ·mol^{-1}) is lower than that for aspect ratio 2.2 (7.85 ± 0.2 kJ·mol^{-1}) and also lower than that of gold nanoprisms (7.94 ± 0.1 kJ·mol^{-1}) (see Figure S1 and Figure S2).

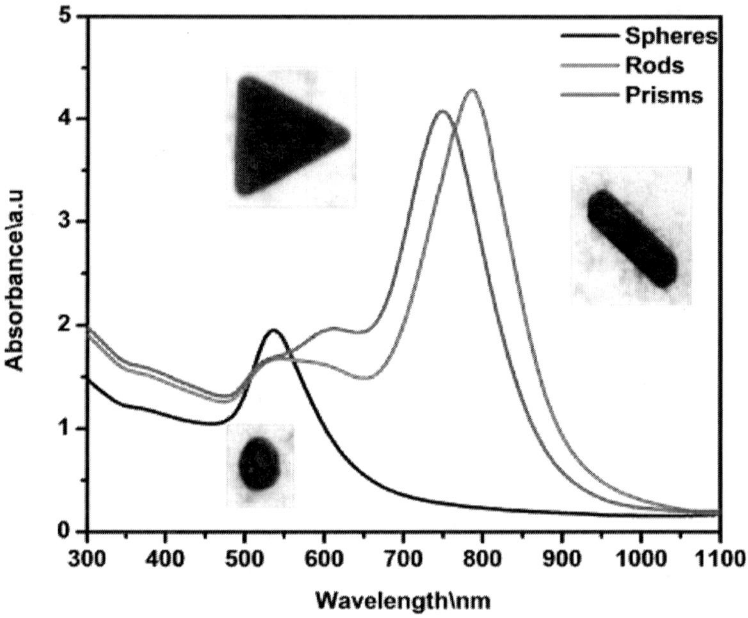

Figure 1. Absorption spectra show the formation of different shapes of gold nanoparticles using different volumes of $AgNO_3$ (0 ml for spheres, 0.1 ml for rods (a.r. = 3.4) and 0.25 for prisms).

Comparing these results with the data previously reported by El-Sayed and coworkers [5] [28] for the thermal stability of gold nanorods prepared with a mixture of two surfactants, it can be observed that the Kraft temperature of the rod shape micelles formed from CTAB (180˚C) is quite higher than that for the rod shape micelles which formed from a mixture of CTAB and TDAB as a co-surfactant (155˚C) because CTAB binds preferentially to the low-energy{111} and {110} crystallographic facets of the gold nanorods than the high-energy {100} crystallographic facets in case of CTAB and TDAB as a co-surfactant [39] .

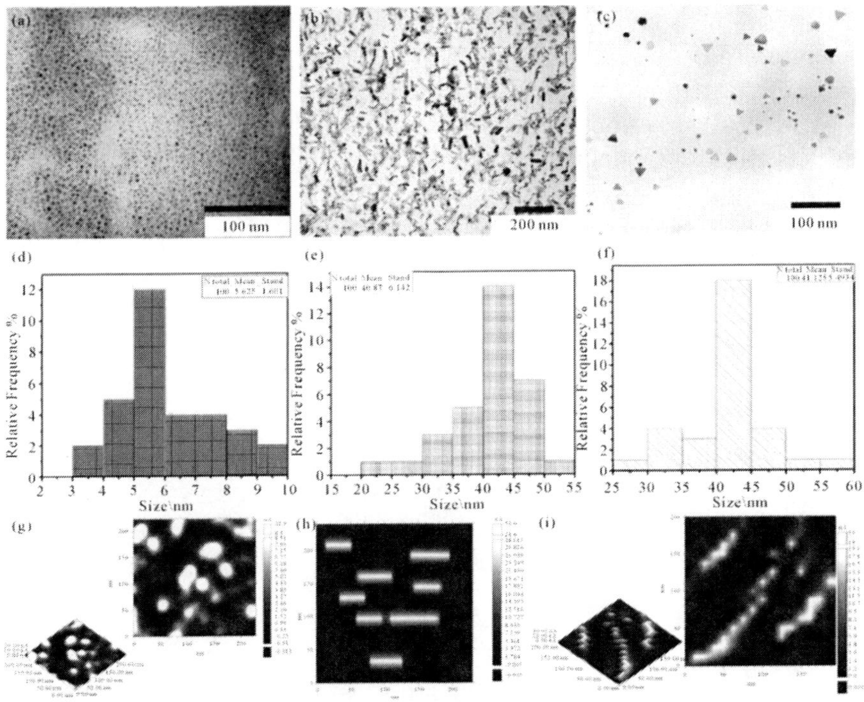

Figure 2. Shows TEM images of the formed gold nanostructures with different shapes: (a)-(c) spheres, rods and prisms respectively using different volumes of $AgNO_3$ (0 ml for spheres, 0.1 ml for rods (a.r. = 3.4) and 0.25 for prisms). The size distribution of the formed particles is shown below the TEM images (d)-(f) and STM images (g)-(i) respectively.

Temperature has different effects on gold nanoprisms than that observed here for the rods (see TEM images in figure 4). For rods, the long rods are not stable, thus, they decompose leading to a decrease in the absorbance and blue shifted of the SP_L band. In case of nanoprisms, the third absorption band characterizing to the nano- prisms disappeared gradually giving absorption spectra similar to the rods. Absorbance and blue shift of the SP_L band took place and the effect of high temperature on nanoprisms can be summarized here in the following steps: 1) Dissolution of the tips of the prism shape which are unstable thermodynamically, giving snapped prisms. 2) These snapped prisms changed on their shape to reach the rod shape particles that are more stable thermodynamically. 3) The bigger particles decomposed leading to a decrease in the absorbance intensity and red shift to lower wavelength, but up to 160°C, particles are still stable.

Finally, the micelles are completely decomposed at 180˚C and the Kraft temperature has found to be ~185˚C (Figure 3(c)). The reason that gold nanoprisms are thermally stable than rods, is due to it has low-energy {111} crystallographic facets, however, gold rods have {110} crystallographic facets with {111} crystallographic facets [39] -[41] .

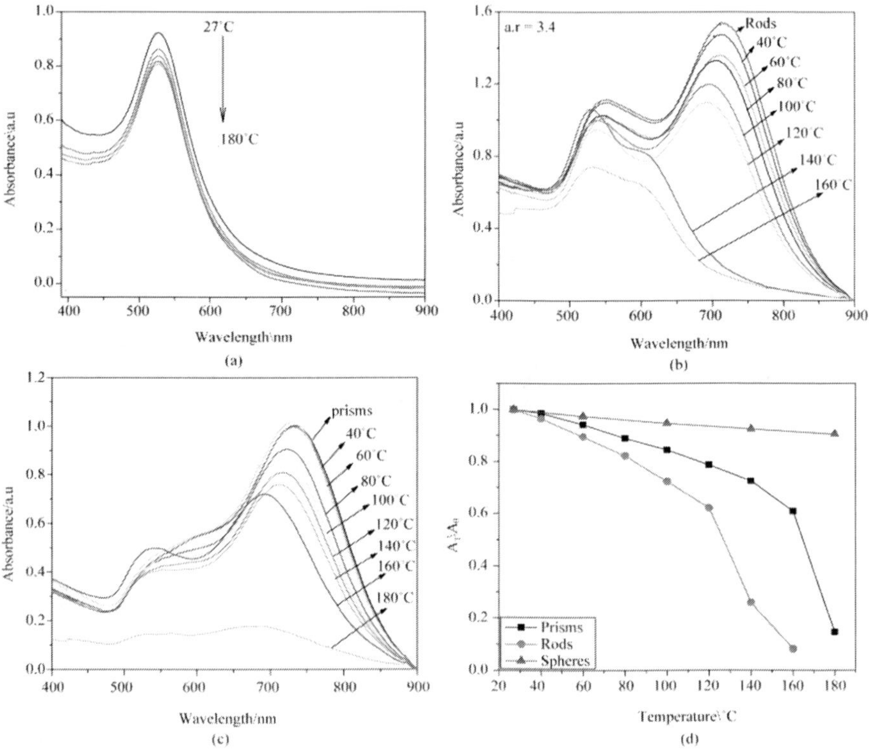

Figure 3. Absorption spectra show the thermal effect on gold nanospheres (a), gold rods of aspect ratio 3.4 (b) gold nano- prisms (c) and normalized absorption spectra of different shapes (d).

3.3. Enhancement of Thermal Stability of Gold Nanoparticles

A solution of PVP is added as another stabilizer to the prepared gold nanoparticles capped with CTAB, and then their thermal stability was studied as mentioned above. Addition of 30% by weight of PVP to the nanorods solution leads to the complete stability of gold nanorods and effectively inhibits the deformation process. Upon addition of PVP, a slight red shift (~4 nm) observed due to changing the dielectric constant of the surrounding medium and the absorbance decreased slightly as a result of decreasing the concentration of rods after addition of PVP. Afterwards, no more absorption shift was observed up to 220˚C (see figure 5(a)). In case of nano- prisms, the particles are completely stable up to 100˚C after addition of 30% PVP, then slight decrease in the absorbance value observed with no shift in the absorption band up to 140˚C. At 160˚C, the intensity of their absorption band decreases remarkably, and the particles aggregated at 180˚C (see figure 5(b)).

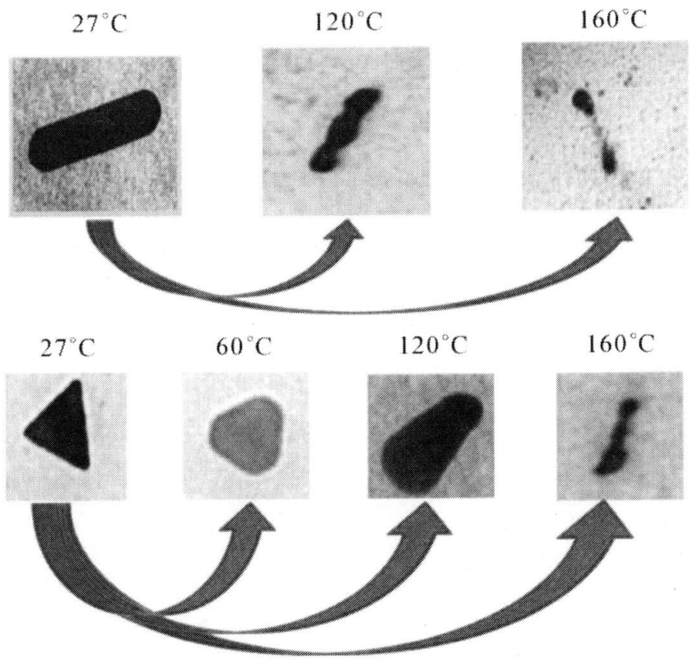

Figure 4. TEM-images showing the decomposition of gold nanorods (aspect ratio 3.4) and gold nanoprisms at different temperatures.

As the symmetry of the shape increases, as the stability decreases (Figure 6). One reason for the stability enhancement could be that the surfactant or polymer normally enhances the dispersivity of nanoparticles in the solution by modification of surface charges. By preventing the agglomeration between the gold nanoparticles, the thermal stability may increase under different temperature compared to the pristine gold nanomaterial. The preferential adsorption of PVP onto {110} faces of rods also plays an important role for the differentiation of distinct facets [42] [43] . Since the shape deformation of the gold nanorods and nanoprisms is affected by the Ostwald ripening, polymer like PVP coating on the nanorparticles as a second nanolayer could effectively increase the barrier of the dissolution or re-deposition processes in the Ostwald ripening, suppressing the shape deformation [33] .

3.4. Hot Carrier Dynamics of Gold Nanoparticles of Different Shapes

The electron dynamics of different shapes gold nanoparticles (prisms, rods, and spheres in figure 1) in solution are studied using a femtosecond pump probe setup which described in details in the experimental section. Single wavelength kinetics of the plasmon bleach recovery of gold nanoparticles of different shapes is recorded [11] . The pump wavelength was 400 nm in all experiments which were performed at room temperature. Figure 7(a) shows the difference in bleach intensity of different shapes of gold nanoparticles with different delay time. The decay curve of the gold nanorods after excitation with 400 nm laser pulse at different delay times and its fitting are shown in figure 7(b).

The power dependence of the relaxation dynamics of gold nanoparticles in hydrogel has been shown [41] . At low pump power, only a fast decay component of a few picoseconds is observed. By increasing the pump power, the fast component of the plasmon bleach recovery becomes longer and a second longer component (~100 ps) appears. The fast decay component is ascribed to the thermal equilibration between the hot electrons and the nanoparticle lattice by electron-phonon (e-ph) interactions while the slow decay component represents the further cooling of the gold nanoparticles by the heat conduction away from the hot lattice to the surrounding medium. The slow decay component has therefore been ascribed to phonon-phonon interactions (ph-ph), which finally leads to the complete cooling of the nanoparticles. The reason for this power dependence of the electron- phonon relaxation time is that the heat capacity of the electron gas is temperature dependent [19] [21] . In addition, higher initial temperatures of the electron gas due to higher excitation powers with a more intense laser pulse will lead to a larger increase of the lattice temperature. If the subsequent removal of the heat by the surrounding medium is the rate-limiting step and depends on the temperature of the lattice, the appearance of second slow relaxation time can be qualitatively understood. More energy dumped into the nanoparticles will simply lead to longer relaxation times of the plasmon bleach because it takes longer for the whole system (gold nanoparticles and the host material) to cool down again.

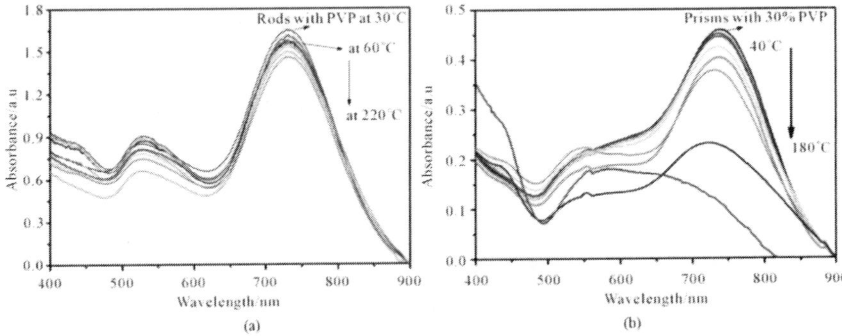

Figure 5. Absorption spectra showing the effect of thermal heating on gold nanorods with 30% PVP from 40˚C to 220˚C. and (b) Absorption spectra showing the effect of the addition of 30% PVP on the thermal stability of the gold nanoprisms.

While especially the electron-phonon coupling has been investigated in the low excitation regime before, it still remains a question, whether the slow phonon-phonon relaxation is mainly due to the heat exchange between the particle and the surrounding medium, or rather is limited by the conduction of the heat away from the particles within the surrounding matrix. Both processes would involve phonon-phonon interactions, but in the first case phonon-phonon coupling between the particles and the matrix is important while the latter would involve mainly the phonons of the surrounding matrix. Comparing the relaxation dynamics of gold nanoparticles of different shapes, it is clear that the relaxation time of the fast component is almost constant (~2.27 ps) in all the three samples. This means that the e-ph coupling has no dependence on the particle shape [8] . On the other hand, the slow component which refers to the ph-ph coupling is shape dependent. The ph-ph coupling is much slower in dots (~176 ps compared with spheres in reference 43 (~150 ps)) than in rods (~142 ps) and prisms (~115 ps) and this could be explained by the fact that nanoprisms have more surface area than that of the rods and the spheres. The Fermi-level also will be different because the presence of semiconducting silver clusters [44] [45] at the tips of rods and prisms that could do electron transfer with gold surface. As shown in our recently work, the concentration of Ag-clusters that used in the formation of prisms is higher than that used in rods, however gold spheres are formed in the absence of Ag-clusters. More studies are performed now to understand the effect of the presence of these clusters on the relaxation dynamics of nanoparticles.

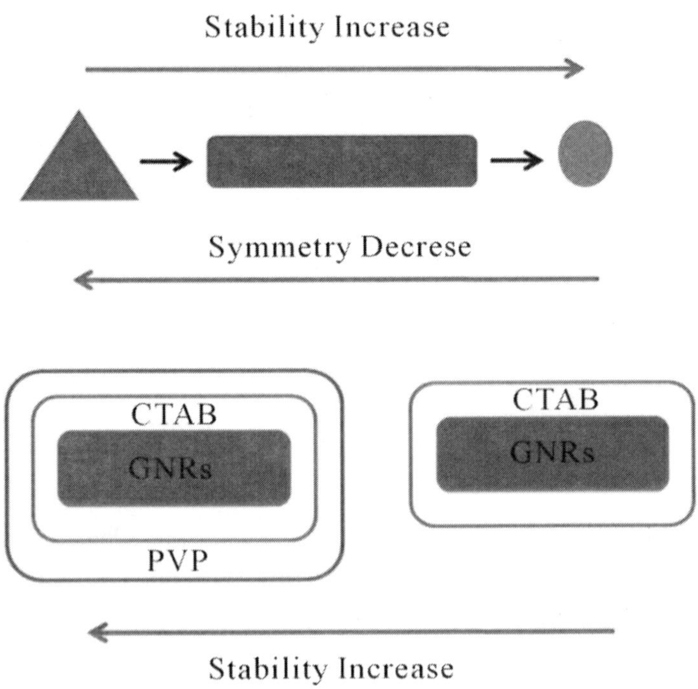

Figure 6. Schematic diagram shows the effect of shape symmetry and capping materials on the thermal stability.

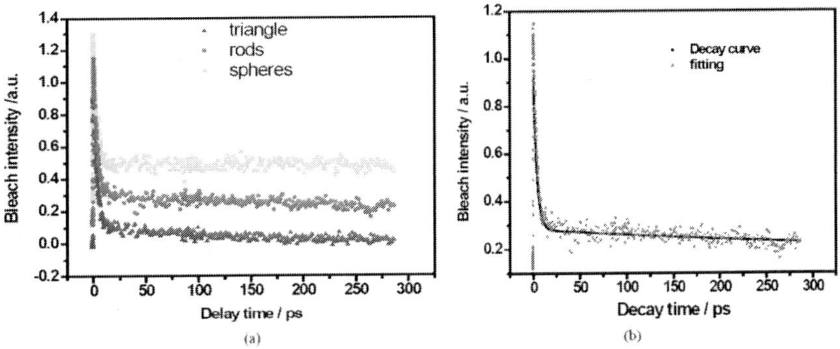

Figure 7. Shows the difference in bleach intensity of different shapes of gold nanoparticles with different delay time (a) and (b) shows the fitting decay curve of the gold nanorods after excitation with 400 nm laser pulse at different delay times.

4. CONCLUSION

We have shown here that the shape and capping materials play an important role in the thermal stability of gold nanostructures. The mechanism of particles dissociation in case of nanoprisms is different than that of nanorods under thermal heating. Great enhancement of the thermal stability has been achieved by adding specific amounts of PVP to the gold nanoparticles. The dynamics of the hot carriers in metallic nanodots, nanorods and nanoprisms induced by femtosecond laser pulses are investigated in gold nanoparticles. It is found that the ph-ph coupling is much slower in dots than in rods and prisms while e-ph coupling is almost the same in these particles. This is due to the large surface area of the nanoprisms compared with the rods or spheres of different thickness and electron transfer between the formed semiconducting Ag clusters and gold surface.

ACKNOWLEDGEMENTS

We acknowledge all the member of NILES-Cairo University specially professor Mona Bakr Mohamed for her support to us.

REFERENCES

1. Murphy, C.J., Sau, T.K., Gole, A.M., Orendorff, C.J., Gao, J., Gou, L., Hun Yadi, S.E. and Li, T. (2005) Anisotropic Metal Nanoparticles: Synthesis, Assembly, and Optical Applications. The Journal of Physical Chemistry B, 109, 13857-13870.

2. Shi, W., Zeng, H., Sahoo, Y., Ohulchansky, T.Y., Ding, Y., Wang, Z.L., Swihart, M. and Prasad, P.N. (2006) A General Approach to Binary and Ternary Hybrid Nanocrystals. NANO Letters, 6, 875-881.

3. Liang, H.-P., Wan, L.-J., Bai, C.-L. and Jiang, L.J. (2005) Gold Hollow Nanospheres: Tunable Surface Plasmon Resonance Controlled by Interior-Cavity Sizes. The Journal of Physical Chemistry B, 109, 7795-7800.

4. Jana, N.R., Gearheart, L. and Murphy, C.J. (2001) Evidence for Seed-Mediated Nucleation in the Chemical Reduction of Gold Salts to Gold Nanoparticles. Chemistry of Materials, 13, 2313-2322.

5. Nikoobakht, B. and El-Sayed, M.A. (2001) Evidence for Bilayer Assembly of Cationic Surfactants on the Surface of Gold Nanorods. Langmuir, 17, 6368-6374.

6. Nikoobakht, B. and El-Sayed, M.A. (2003) Preparation and Growth Mechanism of Gold Nanorods (NRs) Using Seed-Mediated Growth Method. Chemistry of Materials, 15, 1957-1962.

7. El-Sayed, M.A. (2001) Some Interesting Properties of Metals Confined in Time and Nanometer Space of Different Shapes. Accounts of Chemical Research, 34, 257-264.

8. Link, S. and El-Sayed, M.A. (2000) Shape and Size Dependence of Radiative, Non-Radiative and Photothermal Properties of Gold Nanocrystals. International Reviews in Physical Chemistry, 19, 409-453.

9. Link, S. and El-Sayed, M.A. (2003) Optical Properties and Ultrafast Dynamics of Metallic Nanocrystals. Annual Review of Physical Chemistry, 54, 331-366.

10. Link, S., Burda C., Wang Z.L. and El-Sayed, M.A. (1999) Electron Dynamics in Gold and Gold-Silver Alloy Nanoparticles: The Influence of

a Nonequilibrium Electron Distribution and the Size Dependence of the Electron-Phonon Relaxation. The Journal of Chemical Physics, 111, 1255-1264.

11. Mohamed, M.B., Ahmadi, T.S., Link, S., Braun, M. and El-Sayed, M.A. (2001) Hot Electron and Phonon Dynamics of Gold Nanoparticles Embedded in a Gel Matrix. Chemical Physics Letters, 343, 55-63.

12. Link, S., Furube, A., Mohamed, M.B., Asahi, T., Masuhara, H. and El-Sayed, M.A. (2002) Hot Electron Relaxation Dynamics of Gold Nanoparticles Embedded in MgSO4 Powder Compared to Solution: The Effect of the Surrounding Medium. The Journal of Physical Chemistry B, 106, 945-955.

13. Link, S., Hathcock, D.J., Nikoobakht, B. and El-Sayed, M.A. (2003) Medium Effect on the Electron Cooling Dynamics in Gold Nanorods and Truncated Tetrahedra. Advanced Materials, 15, 393-396.

14. Cao, J.G., El-Sayed, A.H., Miller, R.J. and Mantell, D.A. (1998) Femtosecond Photoemission Study of Ultrafast Electron Dynamics in Single-Crystal Au(111) Films. Physical Review B, 58, 10948-10952.

15. Ogawa, S., Nagano, H. and Petek, H. (1998) Stability of a Photoinduced Insulator-Metal Transition in Pr1?xCaxMnO3. Physical Review B, 57, Article ID: R15033. Shahbazyan, T.V. and Perakis, I.E. (2000) Surface Collective Excitations in Ultrafast Pump-Probe Spectroscopy of Metal Nanoparticles. Chemical Physics, 251, 37-49.

16. Hodak, J.K., Martini, I. and Hartland, G.V. (1998) Spectroscopy and Dynamics of Nanometer-Sized Noble Metal Particles. The Journal of Physical Chemistry B, 102, 6958-6967.

17. Bigot, J.Y., Merle, J.C., Cregut, O. and Daunois, A. (1995) Electron Dynamics in Copper Metallic Nanoparticles Probed with Femtosecond Optical Pulses. Physical Review Letters, 75, 4702-4705. Shahbazyan, T.V., Perakis, I.E. and Bigot, J.Y. (1998) Size-Dependent Surface Plasmon Dynamics in Metal Nanoparticles. Physical Review Letters, 81, 3120-3123.

18. Roberti, T.W., Smith, B.A. and Zhang, J.Z. (1995) Ultrafast Electron Dynamics at the Liquid-Metal Interface: Femtosecond Studies Using Surface Plasmons in Aqueous Silver Colloid. The Journal of Chemical Physics, 102, 3860.

19. Faulhaber, A.E., Smith, B.A., Andersen, J.K. and Zhang, J.Z. (1996) Femtosecond Electronic Relaxation Dynamics in Metal Nano-Particles: Effects of Surface and Size Confinement. Molecular Crystals and Liquid Crystals Science and Technology, 283, 25-30.

20. Smith, B.A., Zhang, J.Z., Giebel, U. and Schmid, G. (1997) Direct Probe of Size-Dependent Electronic Relaxation in Single-Sized Au and Nearly Monodisperse Pt Colloidal Nano-Particles. Chemical Physics Letters, 270, 139-144.

21. Del Fatti, N., Flytzanis, C. and Vallee, F. (1999) Ultrafast Induced Electron-Surface Scattering in a Confined Metallic System. Applied Physics B, 68, 433-437.

22. Hodak, J.K., Martini, I. and Hartland, G.V. (1998) Ultrafast Study of Electron-Phonon Coupling in Colloidal Gold Particles. Chemical Physics Letters, 284, 135-141.

23. Smith, B.A., Waters, D.M., Faulhaber, A.E., Kreger, M.A., Roberti, T.W. and Zhang, J.Z. (1997) Preparation and Ultrafast Optical Characterization of Metal and Semiconductor Colloidal Nano-Particles. Journal of Sol-Gel Science and Technology, 9, 125-137.

24. Hodak, J.K., Martini, I. and Hartland, G.V. (2000) Electron-Phonon Coupling Dynamics in Very Small (Between 2 and 8 nm Diameter) Au Nanoparticles. The Journal of Chemical Physics, 112, 5942.

25. Sun, Y. and Xia, Y.J. (2004) Mechanistic Study on the Replacement Reaction between Silver Nanostructures and Chloroauric Acid in Aqueous Medium. Journal of the American Chemical Society, 126, 3892-3901.

26. Mohamed, B.M., Ismail, K.Z., Link, S. and El-Sayed, M.A. (1998) Thermal Reshaping of Gold Nanorods in Micelles. The Journal of Physical Chemistry B, 102, 9370-9374.

27. Petrova, H., Perez Juste, J., Pastoriza-Santos, I., Hartland, G.V., Liz-Marzan, L.M. and Mulvaney, P. (2006) On the Temperature Stability of Gold Nanorods: Comparison between Thermal and Ultrafast Laser-Induced Heating. Physical Chemistry Chemical Physics, 8, 814-821.

28. Veith, G.M., Lupini, A.R., Rashkeev, S., Pennycook, S.J., Mullins, D.R., Schwartz, V., Bridges, C.A. and Dudney, N.J. (2009) Thermal Stability

and Catalytic Activity of Gold Nanoparticles Supported on Silica. Journal of Catalysis, 262, 92-101.

29. Al-Sherbini, M.E. (2010) UV-Visible Light Reshaping of Gold Nanorods. Materials Chemistry and Physics, 121, 349-353.

30. Al-Sherbini, M.E. (2004) Thermal Instability of Gold Nanorods in Micellar Solution of Water/Glycerol Mixtures. Colloids and Surfaces A: Physicochemical and Engineering Aspects, 246, 61-69.

31. Zou, R.X., Zhang, Q., Zhao, Q., Peng, F., Wang, H.J., Yu, H. and Yang, J. (2010) Thermal Stability of Gold Nanorods in an Aqueous Solution. Colloids and Surfaces A: Physicochemical and Engineering Aspects, 372, 177-181.

32. Attia, Y., Buceta, D., Blanco-Varela, C., Mohamed, M., Barone, G. and Lo?pez-Quintela, M.A. (2014) Structure-Directing and High-Efficiency Photocatalytic Hydrogen Production by Ag Clusters. Journal of the American Chemical Society, 136, 1182-1185.

33. Attia, Y., Flores-Arias, M., Nieto, D., Vázquez-Vázquez, C., De La Fuente, G. and López-Quintela, M. (2015) Transformation of Gold Nanorods in Liquid Media Induced by nIR, Visible, and UV Laser Irradiation. The Journal of Physical Chemistry C, 119, 13343-13349.

34. Attia, Y., Buceta, D., Requejo, F., Giovanetti, L. and López-Quintela, M. (2015) Photostability of Gold Nanoparticles with Different Shapes: The Role of Ag Clusters. Nanoscale, 7, 11273-11279.

35. Ledo-Suárez, A., Rivas, J., Rodríguez-Abreu, C.F., Rodríguez, M.J., Pastor, E., Hernández-Creus, A., Oseroff, S.B. and López-Quintela, M.A. (2007) Facile Synthesis of Stable Subnanosized Silver Clusters in Microemulsions. Angewandte Chemie International Edition, 46, 8823-8827.

36. Rodríguez-Vázquez, M.J., Blanco, M.C., Lourido, R., Vázquez-Vázquez, C., Pastor, E., Planes, G.A., Rivas, J. and López-Quíntela, M.A. (2008) Synthesis of Atomic Gold Clusters with Strong Electrocatalytic Activities. Langmuir, 24, 12690-12694.

37. Wang, Z.L., Mohamed, M.B., Link, S. and El-Sayed, M.A. (1999) Crystallographic Facets and Shapes of Gold Nanorods of Different Aspect Ratios. Surface Science, 440, L809-L814.

38. Marzbanrad, E., Rivers, G., Peng, P., Zhaoac, B. and Zhou, N.Y. (2015) How Morphology and Surface Crystal Texture Affect Thermal Stability of a Metallic Nanoparticle: The Case of Silver Nanobelts and Pentagonal Silver Nanowires. Physical Chemistry Chemical Physics, 17, 315-324

39. Wang, Z.L. (2000) Transmission Electron Microscopy of Shape-Controlled Nanocrystals and Their Assemblies. The Journal of Physical Chemistry B, 104, 1153-1175.

40. Chen, J., Wiley, B.J. and Xia, Y. (2007) One-Dimensional Nanostructures of Metals: Large-Scale Synthesis and Some Potential Applications. Langmuir, 23, 4120-4129.

41. Sun, Y., Mayers, B., Herricks, T. and Xia, Y. (2003) Polyol Synthesis of Uniform Silver Nanowires: A Plausible Growth Mechanism and the Supporting Evidence. Nano Letters, 3, 955-960.

42. Chen, W. and Chen, S. (2014) RSC Smart Materials No. 7. RSC, Cambridge.

43. Piñeiro, Y., Rivas, J. and López-Quintela, M.A. (2014) The Emergence of Quantum Confinement in Atomic Quantum Clusters. In: Berti, D. and Palazzo, G., Eds., Colloidal Foundations of Nanoscience, Elsevier, Amsterdam, 81-105.

Chapter 9

Preparation and Characterization of Silica and Clay-Silica Core-Shell Nanoparticles Using Sol-Gel Method

Olfat M. Sadek[1], Safenaz M. Reda[2*], Reem K. Al-Bilali[3]

[1]Department of Materials Science, Institute of Graduate Studies and Research, Alexandria University, Alexandria, Egypt

[2]Chemistry Department, Faculty of Science, Benha University, Benha, Egypt

[3]Chemistry Department, College of Science for Girls, University of Dammam, Dammam, Kingdom of Saudi Arabia

ABSTRACT

Silica and montmorillonite-supported silica nanoparticles were prepared via an acid one step sol-gel process. The synthesized solids were characterized using XRD, FTIR, TEM and N_2 adsorption. The effect of preparing temperatures on the structure and properties of the silica nanoparticles were studied. The results show that the increase of annealing temperature from 25˚C to 200˚C, don't change amorphous state of silica. While for montmorillonite-supported silica the clay platelets are delaminated during the sol-gel process. TEM results showed that the average particle size of silica is increased by increasing temperature due to the particle sintering and the clay-silica nanoparticles possessed core–shell morphology with diameter of 29 nm. The surface area measurements showed that by increasing annealing temperature the surface area was decreased due to aggregation of particle. The clay-silica sample showed lower average pore width than that of the silica prepared at 200°C indicating that it has a macropores structure. The adsorption efficiency of the prepared samples was tested by adsorption of protoporphyrin IX. The highest adsorption efficiency was found for SiO_2 prepared at 200°C. Temkin model describe the equilibrium of adsorption of protoporphyrin IX on caly-silica nanoparticles under different conditions.

KEYWORDS

Silica; Clay-Silica; Sol-Gel Method; Core-Shell; Nanoparticles

1. INTRODUCTION

Silica nanoparticles have attracted wide interest in view of their excellent optical, electrical and thermal properties. It has been applied in many fields, such as catalysis, cosmetics, paints and drug delivery. Different approaches have been used to prepare silica nanoparticles, including the thermal methods, the sol-gel processing technique and water-in-oil (W/O) microemulsion. The sol-gel processing technique for the synthesis of silica nanoparticles has been widely used since pore volume and pore size can be controlled at the chemical solution stage [1-4]. Sol-gel-derived nano silica has many unusual properties such as an adjustable refractive index, a high porosity, a low thermal conductivity, and a low dielectric constant. Therefore, it has become increasingly attractive to use in optical, microelectronic, and thermal applications such as antiglare and antireflective coatings, thermal insulationand other applications [5-8]. Nano structures and properties of silica are greatly dependent on experimental conditions in a sol-gel process [9-12].

Recently, considerable effort has been devoted to the fabrication of core-shell particles, which often exhibit improved physical and chemical properties of silica [4]. The core-shell can be broadly defined as comprising a core (inner material) and a shell (outer layer material). These are able to consist of a wide range of different combinations in close interaction, including inorganic/ inorganic, organic/organic, and inorganic/organic materials which the inorganic/inorganic interaction has utilized in this paper. The core-shell composite particles have been utilized in a wide range of areas, such as electronics, catalysis and separation. The inorganic samples such as clay minerals have wide applications for a long time, such as catalysts and catalytic supports [13], adsorbents of organic substances and clay based porous materials [13]. Since they have enhanced thermal/me chanical stability, high surface area and can effectively reduce the agglomeration of particles [13].

In this paper, the silica nanoparticles have been prepared by an acid one step sol-gel method. The effect of temperatures on the structure of silica is discussed. The effect of addition of sodium montmorillonite clay (Na-MMT) to properties of silica is also observed. Protoporphyrin IX (PPIX) was used as a candidate to study the adsorption efficiency of the prepared samples.

2. EXPERIMENTAL

2.1. Chemicals

All chemicals were used as received without further treatment. Tetraethyl-orthosilicate (acros organics), ethanol, protoporphyrin IX (sigma), montmorillonite (Saudi ceramics company). The molecular structure of protoporphyrin IX is represented in **Figure 1**.

2.2. Preparation of SiO_2 Nanoparticles

SiO_2 nanoparticles were prepared by the acid catalyzed through the hydrolysis of tetraethyl-orthosilicate (TEOS) using the following method: 150 ml of TEOS was added to 155 ml of ethanol and 190 ml of H_2O and stirred thoroughly. The mixture was vigorously stirring for 1 h. 2:3 drops of the hydrochloric acid was added to the reaction mixture as a hydrolysis catalyst to adjust the pH value of alkoxide reaction system. After hydrolysis, the mixture becomes a clear sol. The sols were dried in Teflon dishes at room temperature. The obtained dry xerogels were heated at temperature of 25, 60, 100, 200°C in air for 25 min in order to study the effect of annealing temperature on structure and particle size of silica. These samples are referred to as silica-25, silica-60, silica-100 and silica-200, respectively.

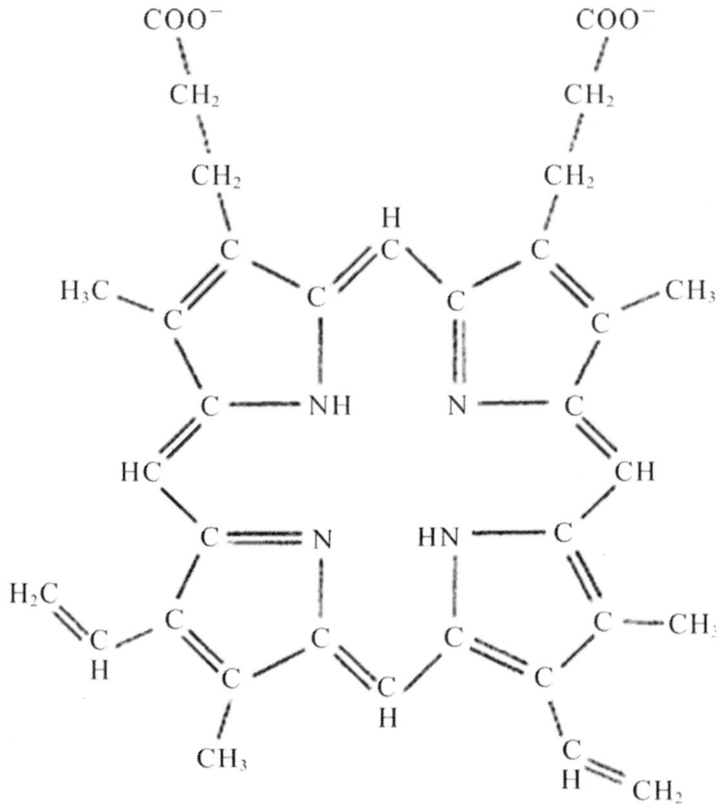

Figure 1. The structure formula of protoporphyrin IX.

2.3. Synthesis of Montmorillonite-Silica Core-Shell

1 g of Na-MMT was dispersed in 19 ml of distilled water to form well-dispersed slurry with magnetic stirring at room temperature for 24 h. the mixture was then stirring ultrasonic at 25 kHz and 100 Watt. 15 ml TEOS and 15 ml ethanol were added into the clay-water slurry. The pH was adjusted to 3 with hydrochloric acid. The admixture was stirred vigorously at 60˚C until a homogenous milk-white gel was formed. The gel was aged for several days to evaporate the solvents for further condensation reaction and then dried in a vacuum at 60˚C for 24 h. A rigid clay-silica was obtained by these treatments was ground into powder for the analysis.

2.4. Characterization

X-ray powder diffraction (XRD) was recorded on Diano Corporation USA diffractometer using Co-radiation (l = 0.179 nm), to study the crystal structure and crystallinity of the SiO_2 nanoparticles. The surface properties namely BET surface area and mean pore radius (r) were determined from N_2adsorption isotherm using conventional or classical volumetric apparatus. Surface Morphology and microstructure was investigated by TEM (Jeol JXA-840 electron probe microanalyzer). The UV-Vis absorption was obtained with a (UVD-2960) labomed spectrophotometer. FT-IR was recorded using Jasco FT/IR spectrometer in wavenumber range from 4000 - 400 cm^{-1}.

2.5 Adsorption of Protoporphyrin IX Molecule on Silica and Clay-Silica Core-Shell

Adsorption of PPIX with the synthesized SiO_2 nanoparticles were carried out as batch tests in 250 ml flasks. 25 ml of PPIX solution with desired concentration is prepared in ethanol as a solvent. The pH of the solution was adjusted using 0.1 N HCl or NaOH solutions. A 0.025 g of nano-SiO_2 powder was then added to the solution, the final silica concentration is 1 g/l. The obtained suspension was immediately shaken for 1 h. After the mixing time elapsed, the suspension was allowed to settle and the supernatant was analyzed using UV-Vis spectrophotometer for the residual dye concentration. The concentration of the remaining dye in the solution was determined by measuring the absorbance intensity at 485 nm [14]. All experiments were performed at room temperature. The investigated ranges of the experimental variables were as follows: dye concentration (10^{-4}, 5×10^{-5} and 1.5×10^{-5} mol/l), pH of solution (2, 10), dosage (1 g/l) and mixing time (1 - 8 days).

3. RESULTS AND DISCUSSION

3.1. Characterization

XRD patterns of the SiO_2, Na-MMT and montmorillonite-silica nanocomposite are shown in **Figure 2**. The results show that there is any characteristic peaks of SiO_2 and the appeared peaks are amorphous in nature. This indicates that the annealing temperature had no significant influence on the phase formation of the SiO_2 nanoparticles till 200°C. The results are in good agreement with that reported in lecture [15]. **Figure 2**(e) shows XRD patterns of the montmorillonite-silica core-shell. The figure shows the broad peak at 22° corresponds to the typical diffraction of amorphous SiO_2 (Si-O short-order structure) [16]. The (001) diffraction peaks of Na-MMT at 6° are vanished,**Figure 2**(f), suggesting that clay platelets are delaminated during the sol-gel process.

FTIR spectra have been used to characterize the prepared samples, **Figure 3**. As shown in this Figure the two broad bands at 3470 and 1639 cm^{-1} are ascribed to the stretching modes and H-O-H bending vibration of the free or absorbed water. Strong absorptions at 1077, 798, 474 cm^{-1}indicate the formation of silica network. The absorption band at 1077 cm^{-1} attributed to Si-O-Si of the SiO_4 tetrahedron and the absorption band at 474 cm^{-1} associated with Si-O-Si or O-Si-O bending mode. These observations afford proof of the generating silica nanoparticles [12].

Figure 3(e) illustrates FT-IR spectra of clay -modified silica. After modification with clay, the bands at about 3470 cm^{-1} corresponds to the stretching vibration of surface hydroxyl groups, and the bands at 956 and 798 cm^{-1} attribute to the -OH bending vibration. The intensity of these characteristic bands of the surface hydroxyl groups increase indicating that the condensation reactions takes place not only between the silica sol particles themselves but also between the sol particles and hydroxyl groups on clay surfaces. A new band at about 545 cm^{-1} appears which is assigned to Al-O-Si of clay. The FT-IR spectra (**Figure 3**(e)) shows no obvious changes of the −OH bands at about 3627, 916 and 847 cm^{-1}. These results suggest that surface hydroxyl groups of the clay were reacted with the hydrolyzed TEOS, and the resultant silica is covalently attached on the clay surfaces [13].

Figure 4 shows TEM images of pure silica and silica-clay samples. The compacted spherical morphology is clearly visible in the TEM images for all samples. Figures 4(a)-(d) illustrate the TEM photographs of silica nanoparticles annealed at different temperatures. The average particle size for silica-25, silica-60, silica-100 and silica-200 was 27, 30, 65, and 202 nm, respectively. As we seen, the average particle size increase by increasing temperatures. The main reason is that the growth of particles by increasing temperature. It is evident that the higher annealing temperature the narrower size distribution was observed. This may be due to the lowered viscosity at higher temperatures, which increase the particle aggregation during precipitation process. The transmission electronic micrographs (TEM) of the obtained clay-silica nanocomposites are shown in**Figure 4**(e). From the insets of this figure, it is evident that the final composite nanoparticles possessed core-shell morphology with diameter of 29 nm was obtained. This is easily understood because more silica network was formed on the clay and surrounding the clay plates [13].

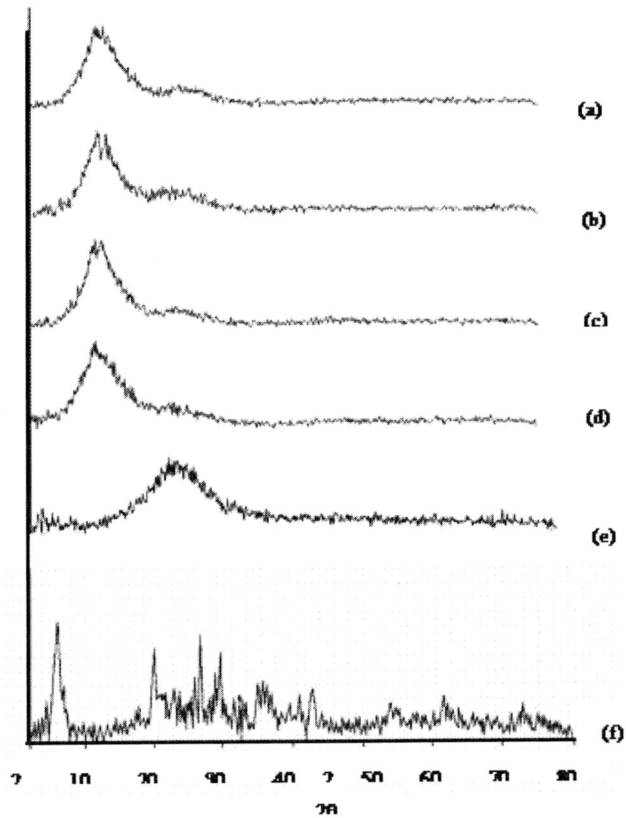

Intensity (a.u)

Figure 2. X-ray diffraction patterns for silica-25 (a), silica-60 (b), silica-100 (c), silica-200 (d), clay-silica nanocomposite (e), and Na-MMT (f).

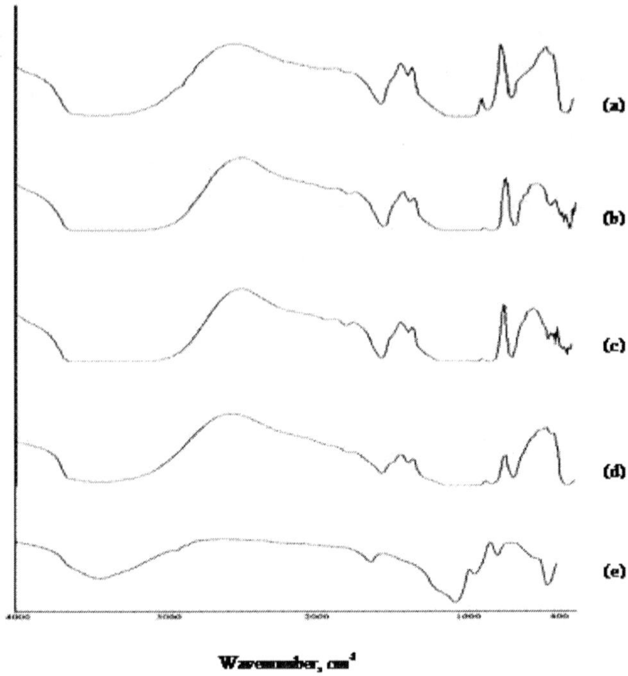

% Transmance (a.u)

Figure 3. FTIR specrta of silica-25 (a), silica-60 (b), silica-100 (c), silica-200 (d) and clay-silica nanocomposite (e).

3.2. Nitrogen Adsorption-Desorption Measurements

The nitrogen adsorption-desorption isotherms are carried out to investigate the surface morphology and physical properties of the silica samples annealed at different temperatures, **Figure 5**. A comparison between the hysteresis loops of these samples indicates that they are of type H4 except for silica-200 for which a superposition of H3 loops is observed. The isotherm of silica-200 (**Figure 5**(d)) presents a type-H3 hysteresis loop ascribed to aggregates (loose assemblages) of plate-like particles forming slit-shaped pores [15]. All of the isotherms are closet in shape to type I because hysteresis occurs during the desorption branches, according to the BDDT classification [15]. However, there are noticeable differences in the shape of their hysteresis loops. However, silica-200 showed shape to type II.

Figure 5(e) shows the adsorption-desorption isotherm for clay-silica core shell. This Figure shows that the clay modified silica present a type-H3 hysteresis loop, which can be attributed to mesoporous structure containing open slit-shaped capillaries with wide bodies and narrow short necks [13].

The BET surface areas (S_{BET}), total pore volumes (V_p) and average pore radius (r) for all examined samples are shown in (**Table 1**). It shows that pristine silica-25 has an obvious increase in surface area compared with the silica prepared at different temperature [15]. The BET surface areas are decreased by increasing annealing temperature for silica samples due to increase aggregation of particle by increasing annealing temperature. The clay silica sample shows higher average pore width than that of the pure silica-200 (as shown in **Table 2**), indicating again macropores and interstices are formed between the silica particles, while mesoporous silica networks are formed in the latter case. Under the acid-catalyzed conditions the sol-gel process involves complete hydrolysis reaction, replacing alkoxide groups ($-OC_2H_5$) with hydroxyl groups (−OH), and subsequent condensation reactions between silanol groups (Si-OH) and hydroxyl groups located on the clay surfaces to form covalent bonds [15]. Meanwhile, the hydrolyzed TEOS also self condense to form large continuous silica networks with mesoporous structure, especially in the case of the high TEOS/clay. As a result, clay platelets are covalently attached with silica, and the growth of the mesoporous networks penetrated into clay interlayer leads to the formation of the exfoliation structure of clay platelets. Measurements of the pore size showed a decreased in pore size by increasing annealing temperature for pure silica reaching the mesoporous region. While for the clay-silica sample has a pore size around 2 nm, the clay-silica sample has a larger amount of pores in the micropore region than in the mesopore region.

3.4. Analysis by t-Plot

The statistical thickness, t, of the physically adsorbed layer as a function of relative pressure is estimated based on the de Boer equation given in equation [15]:

$$t = \left(13.9 / \left(0.034 - \log \frac{P}{P^o} \right) \right)^{1/2} \quad (1)$$

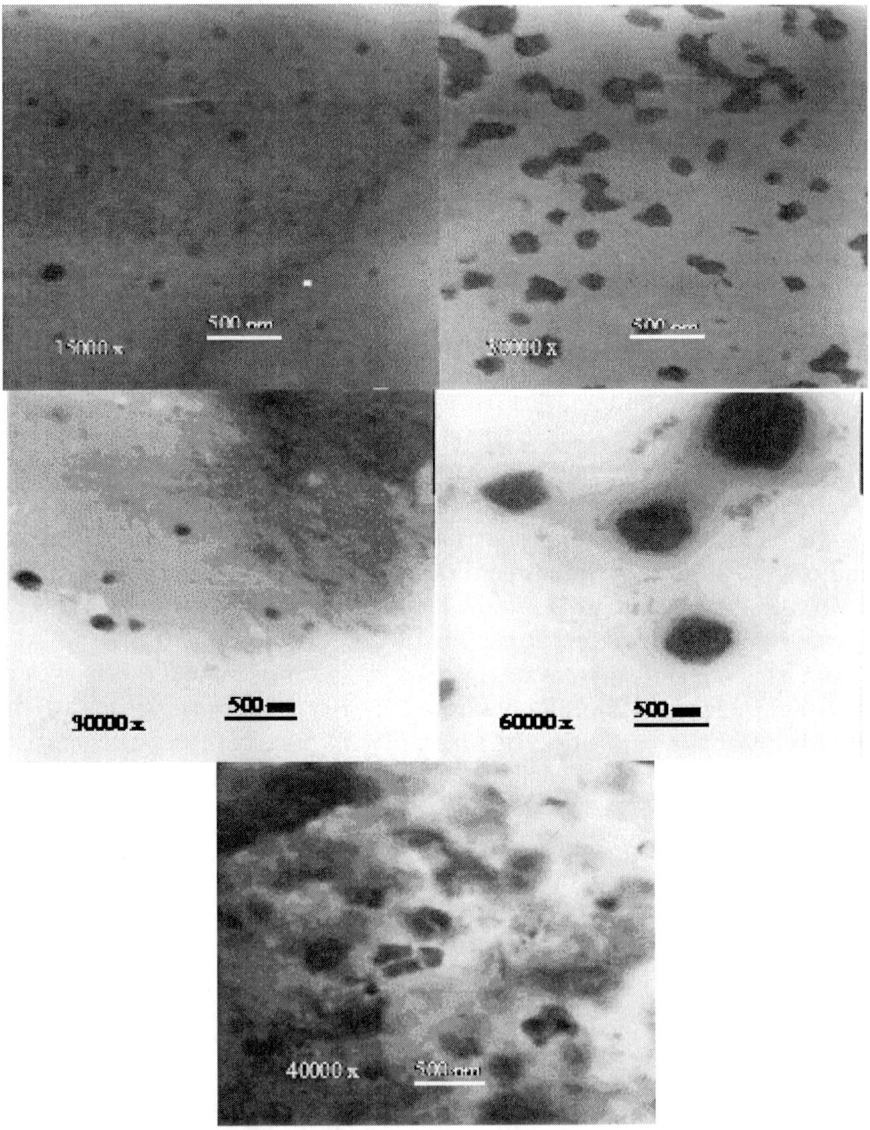

Figure 4. TEM images for silica-25 (a), silica-60 (b), silica-100 (c), silica-200 (d) and clay-silica nanocomposite (e).

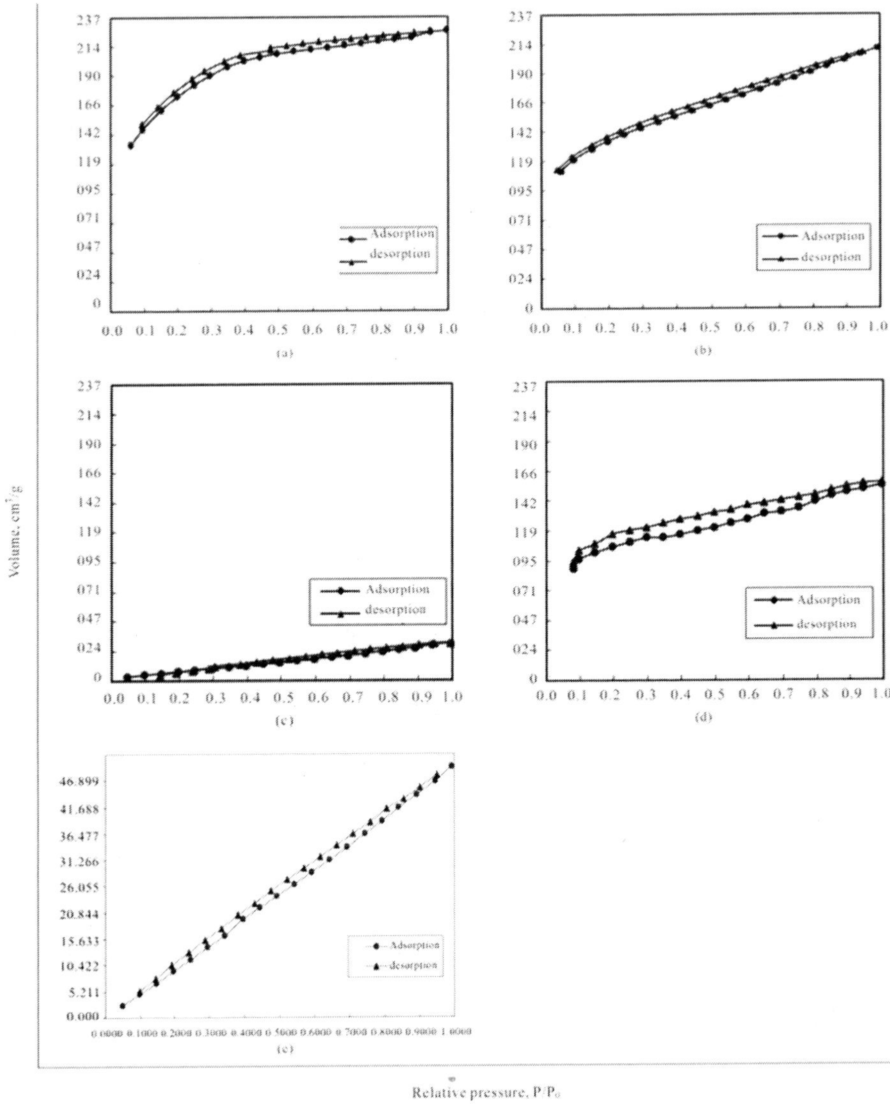

Figure 5. Nitrogen adsorption-desorption isotherm for silica-25 (a), silica-60 (b), silica-100 (c), silica-200 (d) and clay-silica (e).

Table 1. Parameters resulting from the analysis of N_2 adsorption isotherms of the silica samples.

Samples	BET surface area (S_{BET}, m^2/g)	Total pore volume (V_s, cm^3g^{-1})	Average pore width (r, A°)
Silica-25	596.61	0.3498	1.47
Silica-60	451.04	0.3247	1.80
Silica-100	338.32	0.2476	1.83
Silica-200	338.32	0.0466	1.83
Clay-silica core-shell	79.59	0.0763	2.40

Table 2. Parameters resulting from the t-analysis for all investigated samples.

Samples	Pore surface area (S_t, m^2/g)	(S_{ex}, m^2/g)	Microporous surface area (S_{micro}, m^2/g)	Mesoporous surface area (S_{meso}, m^2/g)	Microporous volume (V_{micro}, cm^3g^{-1})	Mesoporous volume (V_{meso}, cm^3g^{-1})
Silica-25	547.51	70.32	526.3	21.21	0.256	0.0937
Silica-60	354.35	102.6	348.4	5.95	0.177	0.1477
Silica-100	145.5	61.73	276.6	1.50	0.1503	0.0973
Silica-200	36.66	28.85	9.622	27.04	0.0042	0.04240
Clay-silica core-shell	66.53	50.97	28.62	37.91	0.0014	0.0749

The volume of gas adsorbed versus t known as t-plot is then used for the pore analysis. **Figure 6**shows two distinct linear stages in the t-plots, initial stage of adsorption on the pore wall; intermediate stage is extrapolated to the adsorption volume ordinate, the positive intercept at the ordinate measures the micropore volume of the sample that have been filled by adsorbed molecules at the beginning of the intermediate stage. Similarly, the linear portion of the last stage measures the external surface area and the total pore volume exclusive of external surface pore volume [15]. Therefore t-plot analysis is more appropriately used for the samples containing both micropores and mesopores [15]. The t-plot pore surface areas of the silica samples and caly-silica, S_t, calculated from the slopes of the initial linear stage of t-plots are similar to the corresponding BET surface areas except that for silica annealed at 100 and 200˚C (**Table 2**). The micropore pore volumes, $V_{PORE,MICROPOROUS}$ is large the corresponding mesopore pore volumes, $V_{PORE,MESOPOROUS}$ (**Table 2**) except silica annealed at 200 and clay-silica samples. This is not surprising as the pore filling of micropores is due to gaseous adsorption that occupies larger surface area than the pore filling of mesopores with liquid-like capillary condensation. **Table 2** shows also that, for all samples external surface S_{ext} areas are less than the corresponding BET surface areas. It can be noted that, if S_t (calculated from the initial slope) is considered as the total surface area arising from all pores, the microporous surface area S_{mic} will be {S_t – ($S_{t,MESOPOROUS}$ + $S_{t,EXTERNAL}$)} that is 526, 348, 276, 9.6 and 28.6 m^2/g for silica prepared at 25, 60, 100, 200 and clay-silica, respectively. The total pore volume of the silica at 25˚C is found to be maximum, 0.349 cm^3/g; and is minimum in case of the silica at 200˚C, 0.0466 cm^3/g. Also, the microspore volume of the silica at 25˚C is found to be the largest, 0.256 cm^3/g. This agrees well with the differences in corresponding S_t surface areas.

3.5. ADSORPTION OF PROTOPORPHYRIN IX MOLECULE ON NANO PARTICLES OF SILICA AND CLAY-SILICA

The adsorption process was carried out using silica nano practices pretreated at 25˚, 60˚, and 200˚C, as well as clay-silica nano particles. It was observed that silica-25 has not ability to absorb the PPIX, there is no change in PPIX concentration in the medium till two days. Other samples: silica-60, silica-100 and silica-200 show slight capability to interact with PPIX. The efficiency of the adsorption process (% eff.) can be determined using the following equation:

$$\% \ eff = \frac{c_o - c_{eq}}{c_o} \times 100 \qquad (2)$$

where c_o and c_{eq} are the concentration of PPIX at initial and equilibrium time respectively, (mg/l).

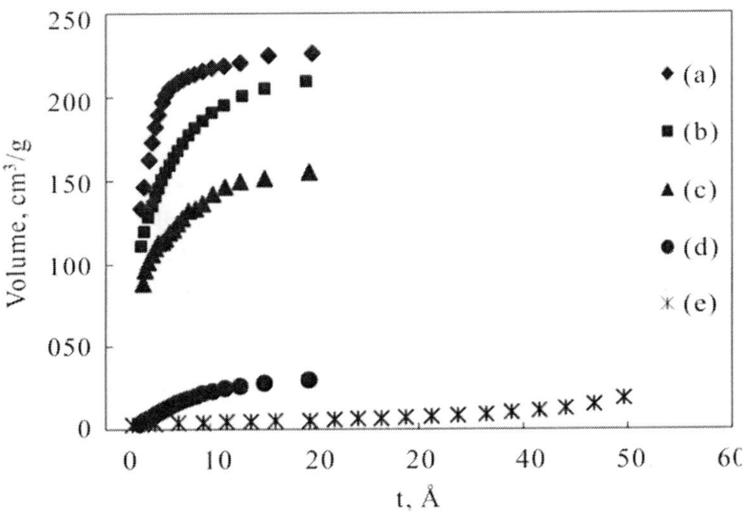

Figure 6. t-analysis for silica-25 (a), silica-60 (b), silica-100 (c), silica-200 (d) and clay-silica (e).

Figure 7 shows the adsorption efficiency of protoporphyrin IX on different SiO_2 surfaces after two days, as function of annealing temperature. Obviously, the silica-200 exhibited the highest adsorption efficiency with 37% in 2days, while silica-100 showed 34% and silica-60 showed 25% adsorption efficiency in the same test period. The silica-25 don't show any change in adsorption in the same time. Also, it was observed that silica-200 show slight increase in its efficiency after the third day to attain 52%, while is no improvement of the efficiency of silica-60 and silica100. The highest adsorption efficiency of silica-200 in may be attributed to its large pore radius (7.6 nm) as observed in BET measurements, indicating that the pore of meso-type considers to the suitable void type permit the interaction of

PPIX molecules with silica surfaces. Also, the surface area of mesopores increased dramatically when the annealing temperature increase to 200°C.

Figure 8 shows the UV-Vis absorption spectra of the PPIX in supernatant solution before and after the adsorption on clay-silica nano particles at pH = 10. After adsorption, the intensity of the sort band at 402 ± 0.5 nm was decreased as expected due to the reduction in the PPIX concentration in solution due to their interaction with clay-silica nano particles.

The initial dye concentration is another important variable that can affect the adsorption process. The effect of initial concentration of protoporphyrin IX at pH = 10 were (10^{-4}, 5×10^{-5} and 1.5×10^{-5} mol/l) was studied on their adsorption onto clay-silica powder under previously determined optimum conditions. The results, in terms of efficiency versus initial concentration of dye, are indicated in **Figure 9**. According to **Figure 9**, efficiency of PPIX increases by increasing dye concentration and doesn't change by increasing concentration up to 1.5×10^{-5} M. This may be attributed to the saturation of adsorption sites on clay-silica by dyes at this concentration. It was noted that the incorporation of clay with silica lead to increase in the mesopores surface area leading to maximize the interaction between PPIX and the claysilica nano particles in point of view their rate and efficiency. We notes that the values of adsorption efficiency of the clay-silica nano particles at pH = 10 is better than the pure silica and clay and reaches to 87% at initial concentration (1×10^{-5} mol/l) of PPIX. This can be explained on the bases of the dissolution of silica network surrounding the clay layers at pH = 10 and the clay plates become available for adsorption of PPIX [15], while in neutral and acidic medium (pH = 2) less clay surface is available to adsorb PPIX, as shown by the proposed structure model (**Figure 10**).

In order to investigate the adsorption equilibrium of the adsorption process, the concentration of PPIX was determined. The adsorbed amount of PPIX (q_t) in mg/g dry sample, on the silica surfaces at any time (t) in min., can the calculated from the equation:

$$q_t = \frac{(c_o - c_t)}{m} v \tag{3}$$

where c_t is the PPIX concentration in the supernatant at any time (mg/l) v is the solution volume (l), and m is the weight of the dray sample (g).

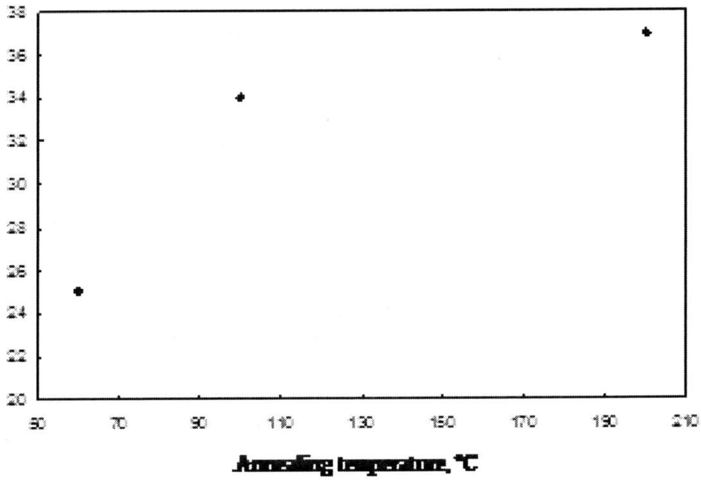

Adsorption efficiency (%)

Figure 7. Dependence of adsorption efficiency of protoporphyrin IX on the annealing temperature of SiO_2 samples after 2 days. (SiO_2 dosage = 1 g/l, initial pH 7, initial dye concentration = 1.5×10^{-5} mol/l).

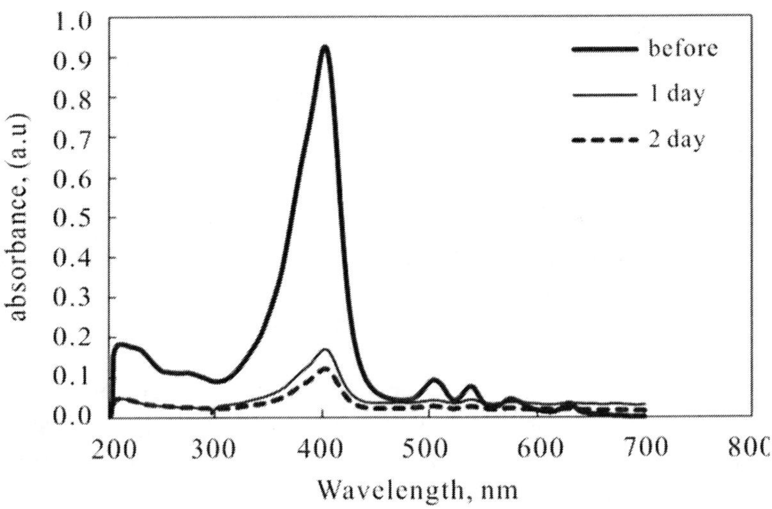

Figure 8. UV-Vis spectra for PPIX before and after adsorption on clay-silica (initial concentration of PPIX = $1 \times 10^{-5\ M,\ pH}$ = 10).

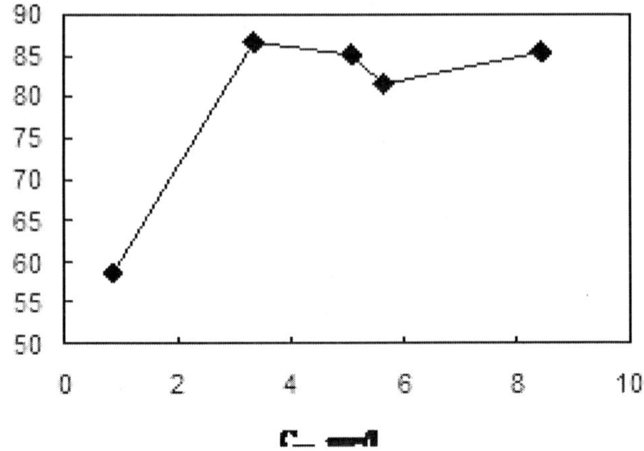

Adsorption efficiency (%)

Figure 9. Effect of concentration on adsorption efficiency of PPIX on clay-silica samples. (clay-silica dosage = 2 g/l and pH = 10).

Figure 11 shows the adsorption isotherm of PPIX onto clay-silica nano particles at pH=10 which was determined by plotting the amount PPIX adsorbed by the nano particles (q_{eq}) in mg/g dray sample versus the equilibrium concentration of PPIX (c_{eq}) in mg/l in solution. The shape of this isotherm is corresponding to s-type, subgroup-3 according to Giles classification [13]. The scurve indicates that there is a solute-solute attractive force of the surface. Also, it indicates that the sorption of PPIX may be inhabited by a competing reaction with the solution, such as a complexation reaction with the surfaces. This means that the PPIX have low affinity to thesurface, but as soon as surface is covered by PPIX molecules, other molecules are adsorbed more easily. There are a second layer can formed readily, thus, it follows up the sub-group 3.

It is very important to establish the most appropriate correlation for the equilibrium data and optimize the adsorption system design to adsorb PPIX on clay-silica nano particles. Four common adsorption equilibrium models (Langmuir, Frundlish, Tembkin and Dubinin-Radushkevish) were tested. The adsorption data were analyzed according to the linear from the previous models, the figures not included, and the correlation coefficient, R^2 was determine which is follow the order:

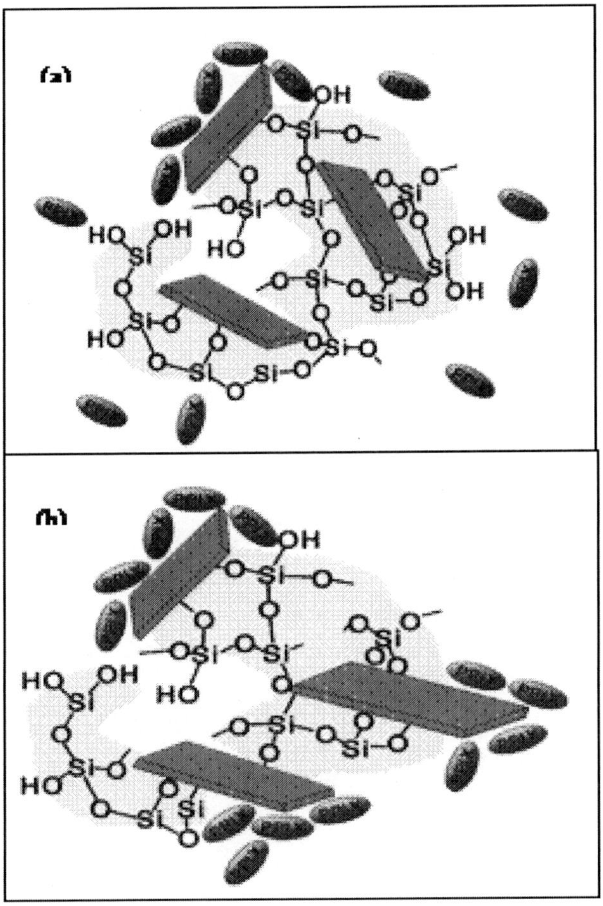

Figure 10. Schematic representation of proposed structure models of clay-silica nanocomposite after adsorption of PPIX (C_o = 10 μM) at (a) neutral medium and pH = 2, and (b) pH = 10.

Temkin (0.88) > Dubinin-Radushkevich (0.76) > Freundlich (0.75) > Langmuir (0.32). According to R^2 values, the most probable models can describe the interaction of PPIX on clay-silica nano particles is Temkin isotherm. It has generally been applied the following form [17].

$$q_{eq} = \beta \ln\left(Ac_{eq}\right) \tag{4}$$

Their linear form can be expressed as:

$$q_{eq} = \beta \ln A + \beta \ln c_{eq} \tag{5}$$

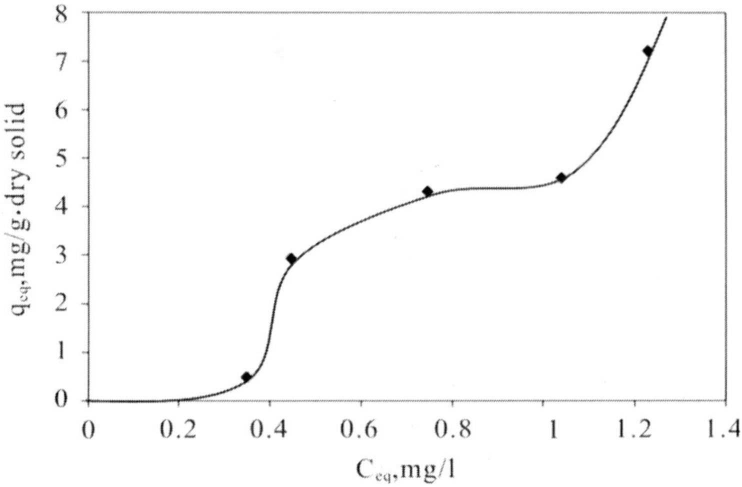

Figure 11. Adsorption isotherm of protoporphrin IX on clay -silica sample. (clay-silica dosage = 1g/l, initial pH = 10).

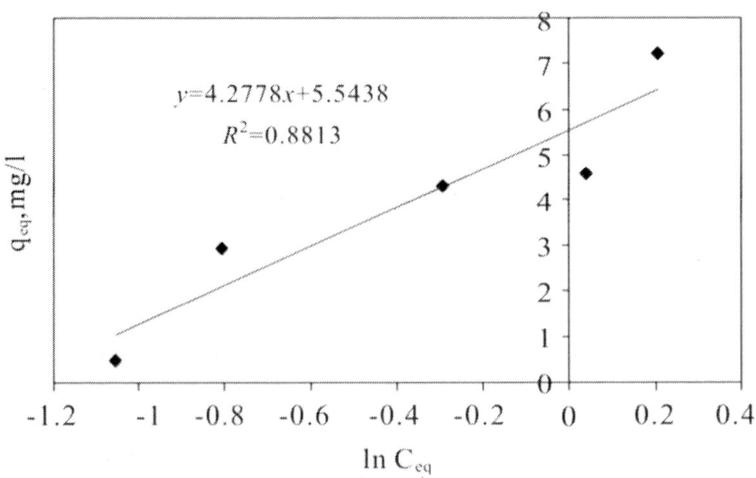

Figure 12. Temkin adsorption model for adsorption of protoporphyrin IX on clay-silica sample at pH = 10.

where β is a constant, refers to the differential surface capacity per unit binding energy, ($\beta = RT/b$), T is absolute temperature, R is the gas constant. β is inversely proportional to the heterogeneity and extent of adsorbate-adsobate interaction. The adsorption data can he analyzed according to Equation 5. Therefore, a plot of q_{eq} versus $\ln c_{eq}$ as shown in **Figure 12**, enable to determine the constants A and β. From the slope of the straight line obtained the value of β can he obtained, β = 4.277, then the value of A can he calculated from the intercept of the line with the coordinate, A = 3.654.

4. CONCLUSION

Silica and clay-silica core-shell nanoparticle were fabricated by one-step catalyzed sol-gel method. The effect of temperature on properties of resulting silica was studied. The prepared samples were characterized using XRD, FT-IR, TEM and N_2 adsorption. The obtained results showed that increasing temperature from 25 to 200°C increase particle size and decease BET surface area due to the particle aggregation. As confirmed from TEM the clay-silica nanoparticles possessed core-shell morphology with diameter of 29 nm. The pore structure properties such as external surface, microporous surface area, mesoporous surface area, total pore volume, microspore volume, mesopore volume are evaluated using t-analysis method. It was found that the microporous surface area is 526, 348, 276, 9.6 and 28.6 m^2/g for silica prepared at 25, 60, 100, 200°C and clay-silica, respectively. The adsorption efficiency of the prepared samples for porotoporphrin IX showed that SiO_2 prepared at 200°C has the highest adsorption efficiency due to its large pore radius (7.6 nm). The results of adsorption showed also that the adsorption of protoporphyrin IX on clay-silica nanoparticles core-shell is better described by the Temkin model.

ACKNOWLEDGEMENTS

The authors very gratefully thank the King Abdulaziz City for Science and Technology for cooperation and support (AT-17-73).

REFERENCES

1. R. Vacassy, R. J. Flatt, H. Hofmann, K. S. Choi and R. K. Singh, “Synthesis of Microporous Silica Spheres,” Journal of Colloid and Interface Science, Vol. 227, No. 2, 2000, pp. 302-315.

2. F. Iskandar and O. K. Mikrajuddin, “In Situ Production of Spherical Silica Particles Containing Self-Organized Mesopores,” Nano Letters, Vol. 1, No. 5, 2001, pp. 231-234.

3. G. De, B. Karmakar and D. Ganguli, “Hydrolysis-Condensation Reactions of TEOS in the Presence of Acetic Acid Leading to the Generation of Glass-Likesilica Microspheres in Solution at Room Temperature,” Journal of Materials Chemistry, Vol. 10, 2000, pp. 2289-2293.

4. J. Zhang, Z. Liu, B. Han, Z. Li, G. Yang, J. Li and J. Chen, “Preparation of Silica and TiO_2-SiO_2 Core-Shell Nanoparticles in Water-in-Oil Microemulsion Using Compressed CO_2 as Reactant and Antisolvent,” Journal of Supercritical Fluids, Vol. 36, No. 3, 2006, pp. 194-201.

5. A. Palaniappan, J. Zhang, X. Su and F. E. H. Tay, “Preparation of Mesoporous Silica Films Using Sol-Gel Process and Argon Plasma Treatment,” Chemical Physics Letters, Vol. 395, No. 1-3, 2004, pp. 70-74.

6. C. Pijolat, J.P. Viricelle, G. Tournier and P. Montmeat, “Application of Membranes and Filtering Films for Gas Sensors Improvements,” Thin Solid Films, Vol. 490, No. 1, 2005, pp. 7-16.

7. M. H. Jo, J. K. Hong, H. H. Park, J. J. Kim, S. H. Hyun and S. Y. Choi, “Application of SiO_2Aerogel Film with Low Dielectric Constant to Intermetal Dielectrics,” Thin Solid Films, Vol. 308-309, 1997, pp. 490-494.

8. B. J. O’sullivan, P. K. Hurley, C. Leveagle and J. H. Das, “Si-SiO_2 Interface Properties Following Rapid Thermal Processing,” Journal of Applied Physics, Vol. 89, No. 7, 2001, pp. 3811-3821.

9. M. Ritala, K. Kukli, A. Rahtu, P. Risnen, M: Leskel, T. Sajavaara and J. Keinonen, “Atomic Layer Deposition of Oxide Thin Films with Metal Alkoxides as Oxygen Sources,” Science, Vol. 288, No. 5464, 2000, pp. 319-321.

10. S. Tanaka, N. Nishiyama, Y. Oku, Y. Egashira and K. Ueyama, "Nano-Architectural Silica Thin Films with Two-Dimensionally Connected Cagelike Pores Synthesized from Vapor Phase," Journal of the American Chemical Society, Vol. 126, No. 15, 2004, pp. 4854-4858.

11. H. Yang, N. Coombs, I. Sokolov and G. A. Ozin, "FreeStanding and Oriented Mesoporous Silica Films Grown at the Air-Water Interface," Nature, Vol. 381, No. 6583, 1996, pp. 589-592.

12. Q. Liu, J. Zhang, Q. Liu, Z. Zhu and J. Chen, "Sol-Gel Synthesis and Characterization of Silica Film with Two Opposite Structures: Nano-Porous and Protuberant," Materials Chemistry and Physics, Vol. 114, No. 1, 2009, pp. 309-312.

13. Z. Qian, G. Hu, S. Zhang and M. Yang, "Preparation and Characterization of Montmorillonite-Silica Nanocomposites: A Sol-Gel Approach to Modifying Clay Surfaces,"' Physica B, Vol. 403, No. 18, 2008, pp. 3231-3238.

14. J. Sun, L. Qiao, S. Sun and G. Wang, "Photocatalytic Degradation of Orange G on Nitrogen-Doped TiO_2 Catalysts under Visible Light and Sunlight Irradiation," Journal of Hazardous Materials, Vol. 155, No. 1-2, 2008, pp. 312-319.

15. P. S. Singh, "High Surface Area Nanoporous Amorphous Silica Prepared by Dodecanol Assisted Silica Formate Sol-Gel Approach," Journal of Colloid and Interface Science, Vol. 325, No. 1, 2008, pp. 207-214.

16. K. M. S. Meera, R. M. Sankar, A. Murali, S. N. Jaisankar and A. B. Mandal, "Sol-Gel Network Silica/Modified Montmorillonite Clay Hybrid Nanocomposites Hydrophobic Surface Coatings," Colloids and Surfaces B: Biointerfaces, Vol. 90, 2012, pp. 204-210.

17. S. K. Parida, S. Dash, S. Patel and B. K. Mishra, "Adsorption of Organic Molecules on Silica Surface," Advances in Colloid and Interface Science, Vol. 121, No. 1-3, 2006, pp. 77-110.

Chapter 10

Synthesis and Physicochemical Characterization of Biodegradable Star-Shaped Poly Lactide-Co-Glycolide-β-Cyclodextrin Copolymer Nanoparticles Containing Albumin

Soodabeh Davaran[1*], Akbar Rezaei[2], Somayeh Alimohammadi[1], Amir Ahmad Khandaghi[3], Kazem Nejati-Koshki[1], Hamid Tayefi Nasrabadi[1], Abolfazl Akbarzadeh[1*]

[1]Department of Medical Nanotechnology, Faculty of Advanced Medical Science, Tabriz University of Medical Sciences, Tabriz, Iran

[2]Department of Science, Ahar Branch, Islamic Azad University, Ahar, Iran

[3]Department of Medicine, Tabriz University of Medical Sciences, Tabriz, Iran

ABSTRACT

The purposes of this research were to synthesize and characterize star-shaped poly lactide-co-glycolide-β-cyclodextrin (PLGA-β-CD) copolymer by reacting L-lactide, glycolide and β-cyclodextrin in the presence of stannous octoate as a catalyst. The structure of PLGA-β-CD copolymer was confirmed with 1H-NMR, 13C-NMR and FT-IR spectra. Albumin as a model peptide drug was encapsulated within nanoparticles made of PLGA-β-CD with a modified double emulsion method. Transmission electron microscopy

(TEM) and scanning electron microscopy (SEM) photomicrographs showed that the nanoparticles have the mean diameter within the range of 80 - 210 nm. Also they were almost spherical in shape. Effects of the experimental parameters, such as copolymer composition, copolymer concentration, and reaction temperature, on particular size and encapsulation efficiency were investigated.

KEYWORDS

Albumin; Star-Shaped Poly Lactide-Co-Glycolide; β-Cyclodextrin; Nanoparticles

1. INTRODUCTION

Therapeutic proteins have been recognized as a novel generation of drugs, mainly due to their highly specific activity [1,2]. Biodegradable nanoparticles as drug delivery systems for proteins or peptides have become an extremely hot topic for decades; these nanoparticles are suitable alternatives for conventional dosage forms [3,4]. As to the use of different biodegradable polymers, PLGA has been extensively used for preparation of nanoparticles due to its safety and physicochemical properties, as it is nontoxic, biodegradable and biocompatible [5,6]. The sequential resistances of this hydrophobic polymeric system are the low entrapment efficiency, the instability of entrapped hydrophilic protein, the burst release and its imperfect release thereafter [7,8]. To overcome demerits of PLGA, nanoparticles described above constructed novel composite nanoparticles, which were composed of biodegradable polymers [9]. Alginate and chitosan were used to form double-walled nanocapsules [10,11]. This kind of nanocapsules not only gives protein a suitable nanoenvironment by dual hydrophilic protection but also makes the initial release retarded [12]. In addition, the dual-walled nanocapsules can be more incorporated in PLGA by W/O emulsification, and the release profiles including the burst release and sustained release can be accommodated with the parameters of alginate, chitosan, and ratios of poly(lactic acid) (PLA) to poly(glycolic acid) (PGA) [13,14]. In addition, the encapsulation efficiency of the composite nanospheres is elevated evidently. Also cyclodextrins (CD) are a group of cyclic oligosaccharides obtained from the enzymatic degradation of starch [15]. They have torus form with hydrophilic outer surface and a lipophilic central hollow space [16]. These properties cause CDs to form inclusion complexes with lipophilic molecules [17,18]. The three most important CDs, α, β and γ, are composed of six, seven or eight D-(+) glucopyranose units, respectively. The aqueous solubility of a protein and drug-CD complex in particular, can be dramatically different from that of the free form. This complexation will affect many of the physicochemical properties of drug without alteration of the essential pharmacokinetic [19,20]. In general, they are used to enhance solubility, bioavailability and stability. Also, they protect drugs against physicochemical and enzymatic degradation [21,22]. No covalent bonds can be between the CD and its guest, so complexation can be considered as a dynamic procedure. Therefore, a drug including within the cyclodextrin hollow may be dissociated upon dilution, displaced by a more appropriate guest or transferred to a matrix for which it has a higher affinity, such as an organic membrane [23]. CDs have been successfully used to modify the release features of polymeric systems mainly due to their facility of forming inclusion complexes with drugs that affect drug solubility and/or diffusivity [24]. Serum albumins are the most abundant proteins in blood plasma. They aid in the transport, supply and metabolism of several endogenous and exogenous ligands [25]. So serum albumin is in use as a model for studying drug-protein interaction in vitro. Many drugs and other bioactive small molecules bind reversibly to albumin and other serum components then utility as carriers [26].

The aim of present work was to assess the merits of PLGA-β-CD nanoparticles as an albumin carrier. For this purpose, the copolymers of PLGA-β-CD were synthesized and characterized. Polymeric nanoparticles were prepared by using a precipitation-solvent evaporation technique. To study the effect of experimental parameters on the properties of these nanoparticls, different formulations of monomers were obtained and fully characterized for morphology, size and encapsulation efficiency.

2. MATERIALS AND METHODS

2.1. Materials

DL-Lactide (LA), glycolide (GA), β-cyclodextrin (β- CD), and stannous octoate were obtained from Sigma Chemical Co. (St. Louis, MO). Dichloromethane (DCM), Polyvinyl alcohol (PVA, Mw 10,000) and solvents were purchased from Merk company. The structure of copolymers was characterized by FT-

IR (Shimadzu 8400, Kyoto, Japan), ^{13}CNMR and ^{1}H-NMR ((Bruker spectra spin 400 MHz, Leipzig, Germany, and their molecular weight and molecular weight distribution (polydispersity index, Mw/Mn) was determined by gel permeation chromatography (GPC, Waters model 515 HPLC pump, Milford, USA). Differential scanning calorimetry (DSC) was conducted in nitrogen atmosphere using a (DSC7 Perkin Elmer, Waltham, USA) calorimeter. Samples were sealed in aluminum pans and heated twice in a nitrogen atmosphere. Drug-loaded polymeric particles were lyophilized by a Christ Alpha 1 - 4 freeze dryer.

2.2. Synthesis of β-CD-PLGA Copolymer

β-CD-PLGA Copolymer nanoparticles of five different ratio (according to table 1) were synthesized Bulk polymerization. LA, GA, and β-CD were heated while stirring to 40˚C, in a nitrogen atmosphere followed by the addition of 0.05% of stannous octoate. The temperature was raised to 150˚C. After 15 min the viscous mixture reacted further at 150˚C for 10 h. The mixture was dissolved in dichloromethane (DCM) as much as possible by boiling and the solvent was separated. The solution was treated with aqueous HCl solution to remove the catalyst. The solution was washed with water and diluted with DMC and added drop wise within half an hour to methanol at –30˚C. At this temperature the mixture was stirred and the product was filtered off and lyophilized. The copolymers are referred to in the text as PLGA-β- CDx, where x designates the mole % of β-CD in the copolymer.

2.3. Preparation of Albumin-Loaded Nanoparticles

Drug-loaded nanoparticles PLGA-β-CD nanoparticles were prepared by using a double emulsion method— evaporation (W/O/W). The albumin was dissolved in HCl (0.1 N) and added to polymer solution (with 10 wt%) in Dichloromethane followed by homogenization in a 20,000 rpm and 60 second primary emulsion W/O was established. These emulsions added to aqueous solution of polyvinyl alcohol (PVA) and for a second time were under Homogenizer. During homogenization the temperature was fixed below 10˚C. After the formation of the emulsion W/O/W solvent was evaporated under vacuum. Nanoparticles obtained by centrifugation at

12,000 rpm for 20 min and were isolated. Nanoparticles supernatants were removed and stored for subsequent experiments.

2.4. Characterization of β-CD-PLGA Copolymer

2.4.1. Fourier Transforms Infrared Spectroscopy

FT-IR spectra obtained by FT-IR spectrophotometer (Shimadzu 8400, Kyoto Japan) for blank and drug loaded nanoparticles using KBr discs.

13CNMR and 1H-NMR spectroscopy The chemical structures of the copolymers were determined by 13 CNMR 1H-NMR (Bruker spectra spin 400 MHz, Leipzig, Germany).

2.4.2. Study of Thermal Properties

Glass transition temperature (Tg) was determined using differential scanning calorimetric measurements (DSC7 Perkin Elemer, Waltham, USA). All measurements were conducted in crimped nonhermetioc aluminium pans by heating the samples at a rate of 10°C/min from 20 to 200°C, with an empty crimped aluminum pan used as reference. Tg was considered at the mid-point temperature of the endothermic drift in the heating curve. All DSC tests were carried out under a 2 ml/min flow of nitrogen to prevent oxidation.

2.4.3. Determine of Average Molecular Weight and Polydispersity Index

The molecular weight and molecular weight distribution (polydispersity index, Mw/Mn) of copolymers were determined by gel permeation chromatography (GPC, waters 515 HPLC pump, USA), that is equipped with a refractive index detector, using a series of high-resolution columns (waters, styragel HR4E 7.8 mm × 300 mm and styragel HR3E 7.8 × 300 mm) and tetrahydrofuran (THF) as a mobile phase. Tetrahydrofuran was used as eluent at 45°C and polystyrene standards (polymer laboratories, Inc. USA). Total time for analyzing of samples which were injected after dissolving in THF solvent was 10 - 15 minutes.

Table 1. Physical properties of star-shaped PLGA–b-CD copolymers.

sample	mole ratio GA:LA	mole ratio CAT:CD:GA:LA	molecular weight	polydispersity	Tg(°C)	Tm(°C)	Yield (%)
PLGA-*β*-CD (I)	0.1:0.3	0.05:0.1:0.25:0.75	12,770	3.3	85	182	65
PLGA-*β*-CD (II)	0.1:0.2	0.05:0.1:0.35:0.65	10,580	3.2	72	155	60
PLGA-*β*-CD(III)	0.1:0.1	0.05:0.1:0.50:0.50	10,058	5.4	48	145	52
PLGA-*β*-CD(IV)	0.2:0.1	0.05:0.1:0.65:0.35	8695	7.2	-	120	40
PLGA-*β*-CD (V)	0.3:0.1	0.05:0.1:0.75:0.25	7225	8.4	-	110	28

2.4.4. Determination of Particle Size and Morphology of the Drug-Loaded Nanoparticle

The morphology of the polymeric nanoparticles was observed using a scanning electron microscopy (SEM). For the scanning electron microscopy (SEM), the lyophilized nanoparticles were placed on a double stick tape over aluminum stubs to get a uniform layer of particles. Samples were then gold-coated using a sputter gold coater. Gold coated particle samples were cooled over liquid nitrogen prior to SEM observations to avoid their melting under high magnification due to the electron beam exposure (Leo Electron Microscopy Ltd., Cambridge, UK). Prior to examination, samples were prepared on aluminum stubs and coated with gold under argon atmosphere by means of a sputter coater.

2.4.5. Determination of the Albumin Entrapment Efficiency

Loaded drug quantity was determined according to the following method after nanoparticles were formed, unbound albumin was separated by centrifugation for 10 min, and the precipitate nanoparticles were separated. The separated nanoparticles was then lyophilized and resulting powder containing the albumin -loaded nanoparticles was dissolved in DMC to obtain a clear solution and analyzed UV spectrophotometrically (Shimadzu UV- 160). Loading capacity was expressed in terms of entrapped drug content, and entrapment efficiency (EE %) as follows:

$$EE\% = \left(\text{entrapped drug content}/\text{initial drug content}\right) \times 100$$

3. RESULT AND DISCUSSION

3.1. Synthesis and Characterization of Star-Shaped PLGA-β-CD Copolymer

Star-shaped polymers because of only one branched center are the simplest branched polymers. These polymers can be produced by chemically linking preformed polymers (polymer first method) or by growing polymer chains from a multifunctional initiator (core first method).

Coupling agents with a compressed dendritic construction have been used to enhance the functionality of the star polymers. Branched polyesters consisting of PVA grafted with chains of PLGA is typical polyester used for pharmaceutical use. In the present study, β-CD was proffered to previously use multifunctional initiators as the potential of CDs and their derivatives as new biodegradable carrier for peptides, proteins, and oligonucleotides with emphasize on the interaction between CDs, these drug candidates, and various endogenous molecules. Chemical structure of star shaped PLGA-β-CD copolymer is shown in **Figure 1**. The copolymer was synthesized by bulk polymerization of the lactones (LA and GA) in the presence of the β-CD using sn-octoate as catalyst. The primary hydroxyl groups of β-CD are expected to begin the polymerization of LA and GA, while it is rather debatable whether the secondary OH will act alike in view of its stereochemical inhibition. The yield of polymer synthesis after precipitation with cold ethers given in **Table 1**. Per mole of β-CD and sn-octoate in all polymerization reactions was 0.1 and 0.05 mole and the variable molar ratios of LA and GA were used. Relatively low amount of β-CD and catalyst was used in order to obtain high molecular weight polymers. **Table 1** and GPC choromarograms in figure 2 shows that increasing the molar ratio of LA to GA, yield of polymerization and the molecular weight of the final polymer increase and the best results in molar ratio of 3 to 1(I) is obtained. This shows the higher activity of LA in copolymerization. Another reactions with variable molar of cyclodextrin were prepared and observed that increasing molar ratio of cyclodextrin growing polymers very minor and the molecular weight of the polymers were low, it related to complexation of monomers in the internal cavity of the cyclodextrin and prevents the initiation of the polymerization reaction. Also, decreasing of cyclodextrin molar ratio from 0.1 mole lead to the formation of linear chains of PLGA. Therefore, the appropriate concentration of cyclodextrin was considered 0.1 moles.

3.2. DSC Measurements

DSC is usually used for study of physical state of nanoparticles. DSC was used to determine the thermal properties of star-shaped copolymers. The expected decrease in the glass transition temperatures Tg and melting points Tm could be observed (**Table 1**). The reduction of interactions between the star-shaped polyester molecules due to their branched structure lowered Tg and Tm compared to linear PLGA copolymers. As shown in **Table 1** increasing lactide in a certain concentration melting point of copolymer were increased becouse lactide increasing number of methyl groups and the amorphous nature of the polymer is more.

3.3. NMR and FT-IR Spectra of Copolymer

The chemical composition of the copolymers was determined with H-NMR and C-NMR by integrating the signals pertaining to each monomer. The freeze dried particles dissolved in DMSO showed the same spectra as that of the original copolymer. This indicated that the composition for the particles was exactly the same as that of the original matrix. Although the ^{1}H NMR spectra (figure 3) of star-branched polyesters show the resonance signals for typical linear PLGA protons (the multiplets at 5.2 and 4.7 ppm related to the LA CH and the GA CH2 respectively, and the singlet at 2.2 ppm correspond to CH3 of LA), there are some other peaks, due to the incorporation of β-CD. The C^2 - C^6 methine and methylene protons of β-CD appeared at 4 - 4.5 ppm the assignment of the peak at 4 ppm to the methylene groups of β-CD was supported by its increase in resonance intensity with an increase in the β-CD molar ratio. The presence of secondary –OH groups at the chain end and/or unreacted –OH groups of polyol could be appear at 4 and 4.3 ppm which may overlap with the peak of methane protons. LA terminal –OH appears at 2 ppm. The signals of the PLGA-β-CD in the ^{13}C NMR spectra (figure 4) were 21.07 ppm LA CH3, 61 ppm LA CH, 173 ppm esteric CO (polylactide-GA ester) 170.07 ppm methylene CH and 17.03 ppm (CH3). The connecting ester bond appeared at about 169 ppm CD-LA ester, and β-CD-CH2-OCO at about 61.2. The FT-IR spectra of these copolymers showed a broad absorption at 2990 cm^{-1} indicating the presence of –OH groups. In addition, an absorption band at 1760 cm^{-1} confirms the formation of –O–C–O ester bonds (figure 5). The GPC traces of the star-shaped PLGA-β-CD copolymers give monodisperse curves. In fact, this is in agreement with previous publications where sn-octoate was used for LA-glycerol or LA pentaerythrol. Therefore, it could be suggested that nitiation of polymerization occurs via the polyol (β-CD)

3.4. Characterization of Made of PLGA-β-CD Copolymer Nanoparticles

Star-shaped PLGA-β-CD(I) copolymer nanoparticles were prepared by a double emulsion-solvent evaporation method in the presence PVA. Figures 6 and 7 shows the SEM and TEM micrographs of the star-shaped PLGA-β-CD nanoparticles. SEM and TEM micrographs of all nanoparticles revealed relatively spherical morphology. The average particle size of samples was 80 - 210 nm.

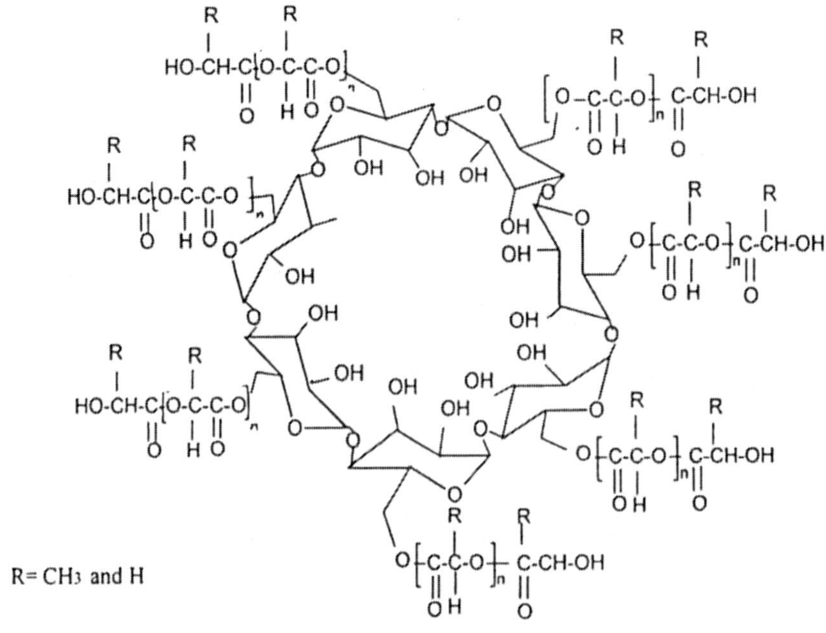

Figure 1. Chemical structure of PLGA-β-CD star-shaped copolymer.

3.5. Effect of Temperature on Yield of Polymerization and Molecular Weight

Polymerizations were conducted in solvent-free and melt conditions. Polymerization performed in 130°C and 150°C and 180°C. Polymerization yield and the molecular weight of the polymers for sample (I) is summarized in **Table 2**. As **Table 2** increasing the temperature of the polymerization reaction from130°C to 180°C the yield of polymerization increased and concentration of unreacted monomers reduced. Also increase in the reaction temperature leads to a decrease in the molecular weight of the polymers. As shown in **Table 2**, between the different temperatures, 150°C is the best temperature for polymerization reaction.

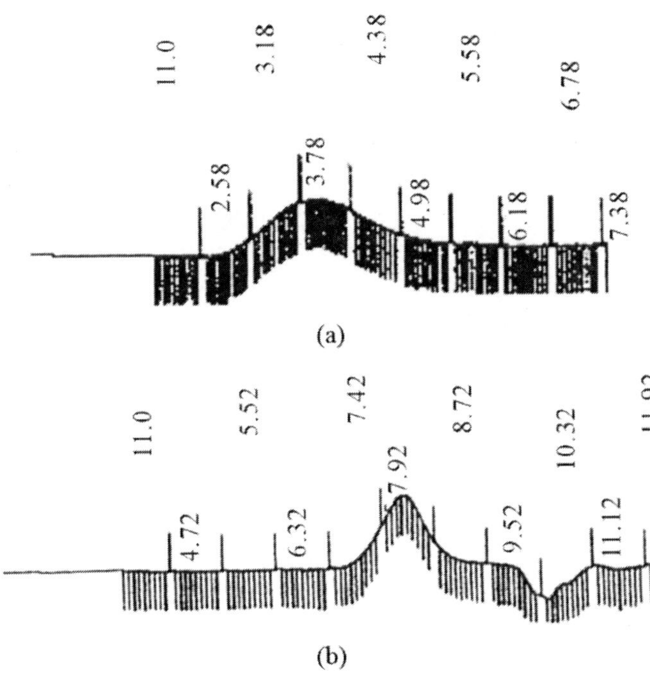

Figure 2. GPC choromarograms of PLGA-β-CD copolymer, a(I) and b (IV).

3.6. Effect of Copolymer Concentration on the Nanoparticles Properties

Concentration of the polymer in the emulsion is an important factor on nanoparticles properties. As shown in tables 3 and 4 polymer molecular weight oil/water ratio and temperature plays an important role in particle size.

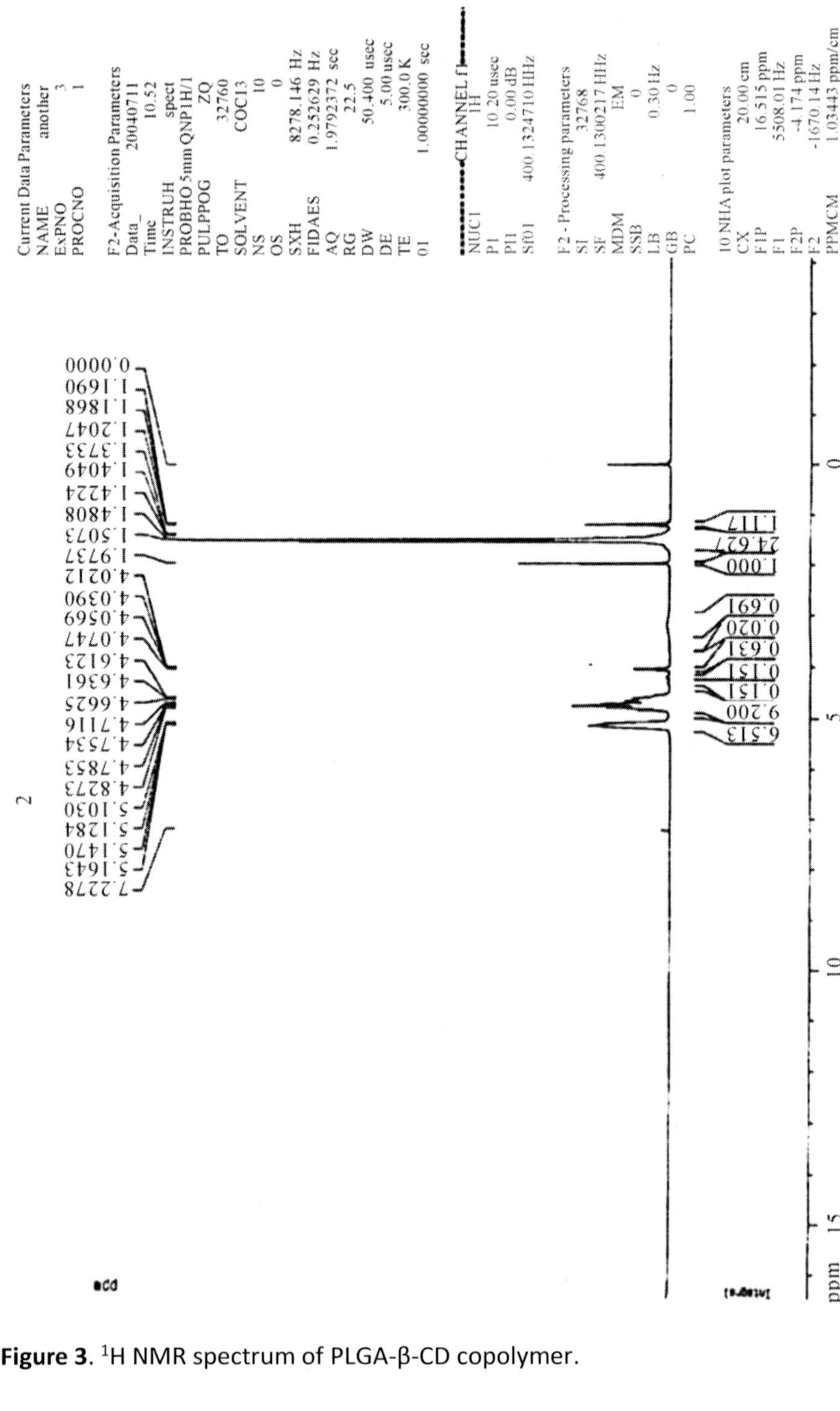

Figure 3. ^{1}H NMR spectrum of PLGA-β-CD copolymer.

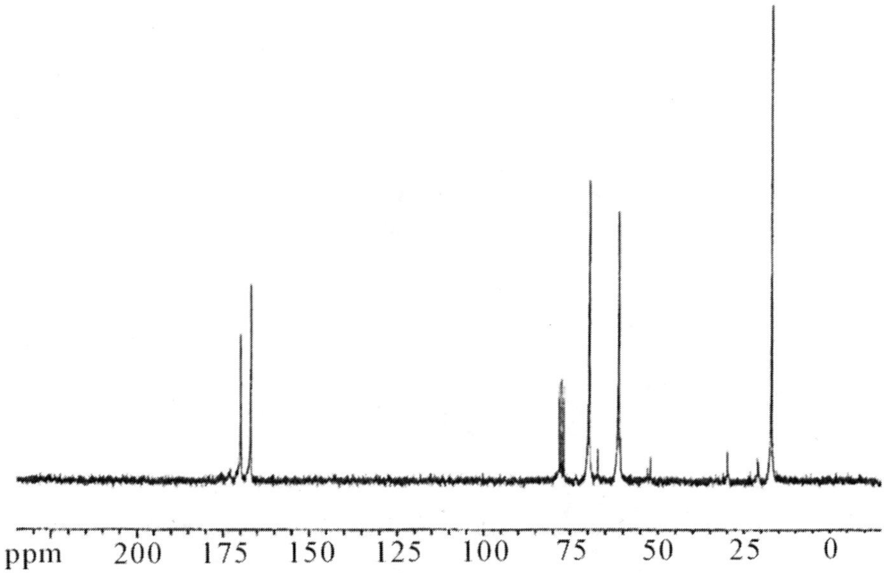

Figure 4. ^{13}C NMR spectrum of PLGA-β-CD copolymer.

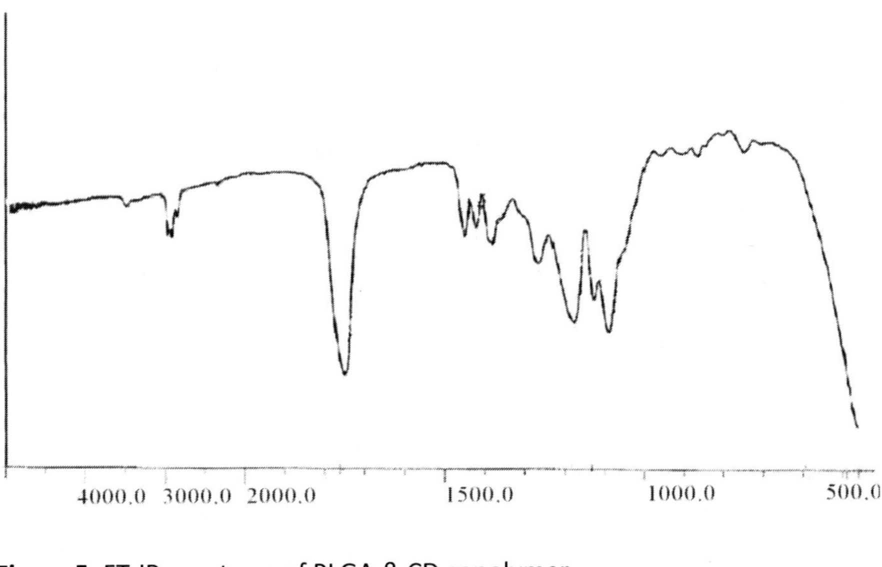

Figure 5. FT-IR spectrum of PLGA-β-CD copolymer.

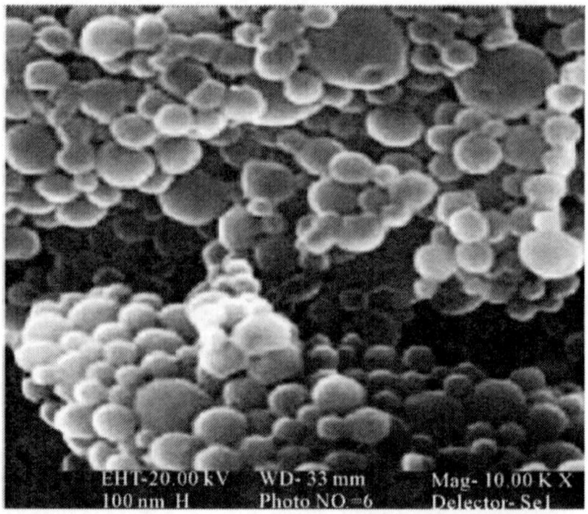

(a)

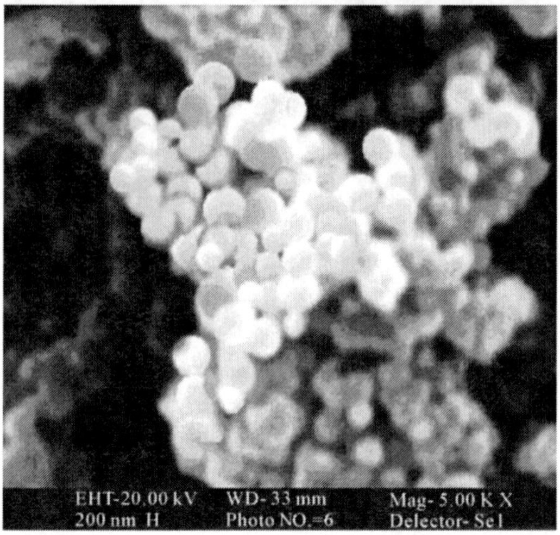

(b)

Figure 6. Scanning electron microscopy (SEM) of nanooparticles surface fabricated by PLGA-β-CD(I) without albumin (a), and PLGA-β-CD(I) with albumin (b).

As shown in **Table 3** increase concentration of polymer microparticle size and efficiency is Mykrvankpsvlasyvn. The reason for this is explained so thick and viscous polymer solution cannot be easily broken into smaller droplets. On the other hand, the formation of a solid shell around Mykrvzrat high concentration of polymer in the polymer layer creates a viscous drop Mykrvzrat faster done. External aqueous phase of the emulsion layer to prevent diffusion of albumin. The high concentration of polymer increases efficiency and reduces nanoparticles initial burst release is albumin.

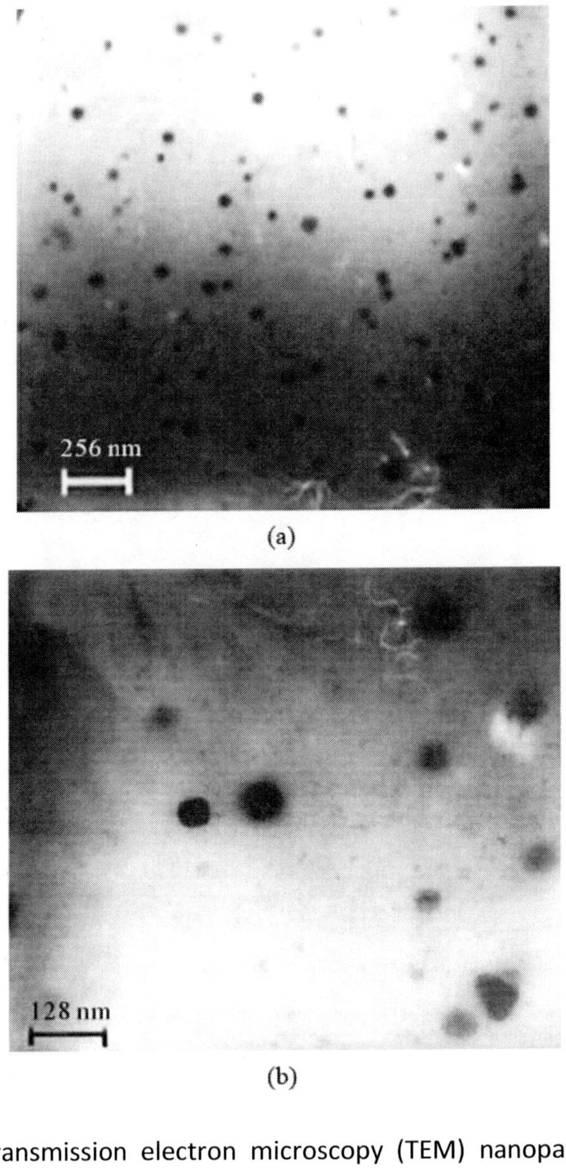

Figure 7. Transmission electron microscopy (TEM) nanoparticles surface fabricated by in 50 mg/ml of PLGA-β-CD copolymer (a), and in 35 mg/ml of PLGA-β-CD copolymer (b).

Table 2. The effect of temperature on the reaction yield and the molecular weight of the final polymer.

polymer	temperature (°C)	Yield (%)	molecular weight
(PLGA-β-CD)I	130	60	11,000
(PLGA-β-CD)I	150	65	12,770
(PLGA-β-CD)I	180	68	9850

Table 3. The effect of polymer concentration on the nanoparticles properties.

polymer (mg/ml)	PVA (mg/ml)	Albumin (%)	encapsulation efficiency (EE %)	Particle Size(nm)
35	0.1	10	45	85 ± 10
50	0.1	10	57.9	100 ± 10
80	0.1	10	65	170 ± 10

Table 4. Effect of aqueous phase volume on particle size and EE.

volume ratio Aqueous: organic	Albumin (%)	encapsulation efficiency (EE %)	Particle Size(nm)
1:15	10	55	150 ± 10
1:25	10	61	95 ± 10
1:45	10	70	80 ± 10

3.7. Effect of Aqueous Phase Volume on Particle Size and EE

According to table 4 increasing in the aqueous phase volume led to a decrease of nanoparticles average size. Coalescence of droplets could be prevented by a large quantity of internal phase. An increase in the internal aqueous could lead to an increase in EE. The Precipitation of the polymer solution phase was accelerated and the hardening time was shortened with the increase of inner phase.

3.8. Effect of Temperature on the Albumin Loaded PLGA-β-CD Copolymers Properties with

Temperemper of preparing nanoparticles known to be a key factor to influence the size and encapsulation efficiency of nanoparticles. As shown in table 5 increasing the temperature from 10°C to 28°C reduce thenanoparticle size becouse the emulsion W/O/W have lower viscosity in high temperatures, so it is easier to break it into smaller droplets.

When the temperature of encapsulation raised to 38°C particles size increased due to the high evaporation rate of the solvent. In addition **Table 5** shows that in 10°C to 38°C encapsulation efficiency is higher than other temperatures. Differences in encapsulation efficiency depend on various parameters such as condition of reaction, nanoparticle size, initial loading, polymer concentration, polymer molecular weight, polymer composition, and special techniques are particle. Formation of a solid layer of polymer around nanoparticle increase efficiency). Nanoparticles which prepared at 38°C quickly become solid and this increased efficiency of encapsulation. The efficiency of prepared nanoparticles in temperatures below and above will be the same [27,28].

4. CONCLUSION

Biodegradable nanoparticles, as drug delivery paradigms, have been extensively used for delivery of wide ranges of small molecules as well as macromolecules such as peptides, proteins and genes [29]. The morphological modification may improve the physicochemical characteristics of the biodegradable polymers. Star-shaped PLGA-β- CD copolymers with different molar ratios of monomers were synthesized and used for the preparation of albumin nanoparticles [30]. In conclusion, the results obtained in the present study showed that star-shaped PLGA-β-CD could be synthesized by ring-opening polymerization of the LA and GA in the presence of β-CD. The albumin loaded nanoparticles could be prepared by double emulsion technique. It was found that the copolymers' composition influences the mean geometric diameter of the albumin loaded nanoparticles, as well as their loading efficiency. These results showed that PLGA-β-CD nanoparticles could be an effective carrier for anticancer drug delivery. An increase in the internal aqueous phase volume led to a decrease in particles' average size. The high entrapment efficiency (EE) (about 70%) was obtained for 80 nm particles [31,32].

Table 5. Effect of temperature on the albumin loaded PLGA-β-CD copolymers properties.

Temperature (°C)	Albumin (%)	encapsulation efficiency (EE %)	Particle Size(nm)
10	10	69	170
22	10	52	100
28	10	58	85
33	10	63	125
38	10	67	150

COMPETING INTERESTS

The authors declare that they have no competing interests.

AUTHORS' CONTRIBUTIONS

KN conceived of the study and participated in its design and coordination. AA and SD participated in the sequence alignment and drafted the manuscript. All authors read and approved the final manuscript.

ACKNOWLEDGEMENTS

The authors thank Department of Medical Nanotechnology, Faculty of Advanced Medical Science of Tabriz University, and Department of Science, Ahar Branch, Islamic Azad University, Ahar, Iran for all supports provided.

REFERENCES

1. S. P. Schwederman, M. Cardamone, M. R. Brandon, A. Klibanov and R. Langer, "Stability of Proteins and Their Delivery from Biodegradable Microspheres," In: S. Cohen and H. Bernstein, Eds., Microparticulate Systems for the Delivery of Proteins and Vaccines, Marcel Dekker, New York, 1996, pp. 1-50.

2. D. Blanco and M. J. Alonso, "Protein Encapsulation and Release from Poly(lactide-co-glycolide) Microspheres: Effect of the Protein and Polymer Properties and of the Co-Encapsulation of Surfactants," European Journal of Pharmaceutics and Biopharmaceutics, Vol. 45, 1998, pp. 285-294.

3. S. Alimohammadi, R. Salehi, N. Amini and S. Davaran, "Synthesis and Physicochemical Characterization of Biodegradable PLGA-Based Magnetic Nanoparticles Containing Amoxicilin," Bulletin of the Korean Chemical Society, Vol. 33, No. 10, 2012, pp. 3225-3232. http://dx.doi.org/10.5012/bkcs.2012.33.10.3225

4. C.-H. Zheng, J.-Q. Gao, Y.-P. Zhang and W.-Q. Liang, "A Protein Delivery System: Biodegradable AlginateChitosan-Poly(lactic-co-glycolic acid) Composite Microspheres," Biochemical and Biophysical Research Communications, Vol. 323, No. 4, 2004, pp. 1321-1327. http://cdx.doi.org/10.1016/j.bbrc.2004.09.007

5. H. B. Ravivarapu, K. Burton and P. K. Deluca, "Polymer and Microsphere Blending to alter the Release of a Peptide from PLGA Microspheres," European Journal of Pharmaceutics and Biopharmaceutics, Vol. 50, No. 2, 2000, pp. 263-270. http://dx.doi.org/10.1016/S0939-6411(00)00099-0

6. J. K. Li, N. Wang and X. S. Wu, "A Novel Biodegradable System Based on Gelatin Nanoparticles and Poly(lacticco-glycolic Acid) Microspheres for Protein and Peptide Drug Delivery," Pharmacological Research, Vol. 86, 1997, pp. 891-895.

7. M. Diwan and T. G. Park, "Pegylation Enhances Protein Stability Stability during Encapsulation in PLGA Microspheres," Journal of Controlled Release, Vol. 73, No. 2-3, 2001, pp. 233-244. http://dx.doi.org/10.1016/S0168-3659(01)00292-9

8. F. Karen, K. Griebenow, L. Hsieh, A. M. Klibanov and R. Langer, "FTIR Characterization of the Secondary Structure of Proteins Encapsulated with PLGA Microspheres," Journal of Controlled Release, Vol. 58, No. 3, 1999, pp. 357-366.

9. J. Kang and S. P. Schwendeman, "Comparison of the Effects of $Mg(OH)_2$ and Sucrose on the Stability of Bovine Serum Albumin Encapsulated in Injectable Poly(d,llactide-co-glycolide) Implants," Biomaterials, Vol. 23, No. 1, 2002, pp. 239-245.http://dx.doi.org/10.1016/S0142-9612(01) 00101-6

10. M. L. Gonzalez-Rodrıguez, M. A. Holgado, C. SanchezLafuente, A. M. Rabasco and A. Fini, "Alginate/Chitosan Particulate Systems for Sodium Diclofenac Release," International Journal of Pharmaceutics, Vol. 232, No. 1-2, 2002, pp. 225-234.http://dx.doi.org/10.1016/S0378-5173(01) 00915-2

11. X. Z. Shu and K. J. Zhu, "The Release Behavior of Brilliant Blue from Calcium-Alginate Gel Beads Coated by Chitosan: The Preparation Method Effect," European Journal of Pharmaceutics and Biopharmaceutics, Vol. 53, No. 2, 2002, pp. 193-201.http:// dx.doi.org/10.1016/S0939-6411(01)00247-8

12. D. W. Lee, S. J. Hwang, J. B. Park and H. J. Park, "Preparation and Release Characteristics of Polymer-Coated and Blended Alginate Microspheres," Journal of Microencapsulation, Vol. 20, 2003, pp. 179-192.

13. M. Diwan and T. G. Park, "Stabilization of Recombinant Interferon-A by Pegylation for Encapsulation in PLGA Microspheres," International Journal of Pharmaceutics, Vol. 252, No. 1-2, 2003, pp. 111-122. http://dx.doi.org/10.1016/S0378-5173(02)00636-1

14. M. D. Blanco and M. J. Alonso, "Development and Characterization of Protein-Loaded Poly(lactide-co-glycolide) Nanospheres," European Journal of Pharmaceutics and Biopharmaceutics, Vol. 43, No. 3, 1997, pp. 287-294. http://dx.doi.org/10.1016/S0939-6411(97)00056-8

15. F. Lallemand, O. Felt-Baeyens, K. Besseghir, F. BeharCohen and R. Gurny, "Cyclosporine Adelivery to the Eye: A Pharmaceutical Challenge," European Journal of Pharmaceutics and Biopharmaceutics, Vol. 56, No. 3, 2003, pp. 307-318. http://dx.doi.org/10.1016/S0939-6411(03)00138-3

16. T. Loftsson, A. Magnudottir, M. Masson and J. F. Sigurjonsdottir, "Self-Association and Cyclodextrin Solubilization of Drugs," Journal of Pharmaceutical Sciences, Vol. 91, No. 11, 2002, pp. 2307-2316. http://dx.doi.org/10.1002/jps.10226

17. T. Loftsson, K. Matthiasson and M. Masson, "The Effect of Organic Salts on Cyclodextrin Solubilization of Drugs," International Journal of Pharmaceutics, Vol. 262, No. 1-2, 2003, pp. 101-107. http://dx.doi.org/10.1016/S0378-5173(03)00334-X

18. G. Orive, R. M. Hernandez, A. R. Gascon and J. L. Pedraz, "Micro and Nano Drug Delivery Systems in Cancer Therapy," Cancer Therapy, Vol. 3, 2005, pp. 131-138.

19. Y. P. Li, Y. Pei, X. Zhang, Z. Gu, Z. Zhou, W. Yuan, J. Zhou, J. Zhu and X. Gao, "PEGylated PLGA Nanoparticles as Protein Carriers, Synthesis, Preparation and Biodistribution in Rats," Journal of Controlled Release, Vol. 71, No. 2, 2001, pp. 203-211.http://dx.doi.org/10.1016/S0168-3659(01)00218-8

20. A. Ahmadi, H. Shirazi, N. Pourbagher, A. Akbarzadeh and K. Omidfar, "An Electrochemical Immunosensor for Digoxin Using Core-Shell Gold Coated Magnetic Nanoparticles as Labels," Molecular Biology Reports, 2014, Epub Ahead of Print.

21. K. Nejati-Koshki, A. Akbarzadeh, M. Pourhasan-Moghadam, A. Abhari and H. Dariushnejad, "Inhibition of Leptin and Leptin Receptor Gene

Expression by Silibinin-Curcumin Combination," Asian Pacific Journal of Cancer Prevention, Vol. 14, No. 11, 2013, pp. 6595-6599.

22. M. Pourhassan-Moghaddam, M. Rahmati-Yamchi, A. Akbarzadeh, H. Daraee, K. Nejati-Koshki, Y. Hanifehpour and S. W. Joo, "Protein Detection through Different Platforms of Immuno-Loop-Mediated Isothermal Amplification," Nanoscale Research Letters, Vol. 8, No. 8, 2013, p. 485.

23. M. Mollazade, K. Nejati-Koshki, A. Akbarzadeh, Y. Hanifehpour, N. Zarghami and S. W. Joo, "PAMAM Dendrimers Arugment Inhibitory Effect of Curcumin on Cancer Cell Proliferation: Possible Inhibition of Telomerase," Asian Pacific Journal of Cancer Prevention, Vol. 14, No. 11, 2013, pp. 6925-6928. http://dx.doi.org/ 10.7314/ APJCP. 2013. 14.11.6925

24. R. Rezaei-Sadabady, N. Zarghami, A. Barzegar, A. Eidi, A. Akbarzadeh and M. Rezaei-Tavirani, "Studies of the Relationship between Structure and Antioxidant Activity in Interesting Systems, Including Tyrosol, Hydroxytyrosol Derivatives Indicated by Quantum Chemical Calculations," Soft, Vol. 2, No. 2, 2013, pp. 13-18.

25. Y. J. Hu, Y. Liu, Z. B. Pi and S. S. Qu, "Interaction of Cromolyn Sodium with Human Serum Albumin: A Fluorescence Quenching Study," Bioorganic & Medicinal Chemistry, Vol. 13, No. 24, 2005, pp. 6609-6614. http://dx.doi.org/10.1016/j.bmc.2005.07.039

26. E. Alarcon, A. M. Edwards, A. Aspee, F. E. Moran, C. D. Borsarelli, E. A. Lissi, D. Gonzalez-Nilo, H. Poblete and J. C. Scaiano, "Photophysics and Photochemistry of Dyes Bound to Human Serum Albumin Are Determined by the Dye Localization," Photochemical & Photobiological Sciences, Vol. 9, No. 1, 2010, pp. 93-102.http://dx.doi.org/ 10.1039/b9pp00091g

27. A. Valizadeh, H. Mikaeili, M. Samiei, S. M. Farkhani, N. Zarghami, M. Kouhi, A. Akbarzadeh and S. Davaran, "Quantum Dots: Synthesis, Bioapplications, and Toxicitynanoscale," Research Letters, Vol. 7, 2012, p. 480.

28. A. Akbarzadeh, M. Samiei, S. W. Joo, M. Anzaby, Y. Hanifehpour, H. T. Nasrabadi and S. Davaran, "Synthesis, Characterization and in Vitro Studies of AdriamaycinEncapsulated Magnetic Nanoparticles Grafted to Smart Copolymers on A549 Lung Cancer Cell Line," Journal of

Nanobiotechnology, Vol. 10, No. 1, 2012, p. 46.http:// dx.doi.org/10.1186/1477-3155-10-46

29. A. Akbarzadeh, R. Rezaei-Sadabady, S. Davaran, S. W. Joo, N. Zarghami, Y. Hanifehpour, M. Samiei, M. Kouhi and K. Nejati-Koshki, "Liposome: Classification, Preparation, and Applications," Nanoscale Research Letters, Vol. 8, No. 1, 2013, p. 102.http://dx.doi.org/10.1186/1556-276X-8-102

30. A. Akbarzadeh, N. Zarghami, H. Mikaeili, D. Asgari, A. M. Goganian, K. Khaksar, S. Mohammad and D. Soodabeh, "Synthesis, Characterization and in Vitro Evaluation of Novel Polymer-Coated Magnetic Nanoparticles for Controlled Delivery of Adriamaycin," Nanotechnology, Science and Applications, Vol. 5, 2012, pp. 1-13.

31. A. Akbarzadeh, H. Mikaeili, N. Zarghami, R. Mohammad, A. Bsrkhordari abd S. Davaran, "Preparation and In-Vitro Evaluation of Adriamaycin-Encapsulated Fe_3O_4 Magnetic Nanoparticles Modified with Biocompatible Copolymer," International Journal of Nanomedicine, Vol. 7, 2012, pp. 1-16.

32. A. Akbarzadeh, M. Samiei and S. Davaran, "Magnetic Nanoparticles: Preparation, Physical Properties, and Applications in Biomedicine," Nanoscale Research Letters, Vol. 7, No. 1, 2012, p. 144. http://dx.doi.org/10.1186/1556-276X-7-144

Chapter 11

Preparation and Characterization of Novel Drug-Inserted-Montmorillonite Chitosan Carriers for Ocular Drug Delivery

Dongzhi Hou[1*], Ruyi Gui[1], Sheng Hu[1], Yi Huang[1], Zuyong Feng[2*], Qineng Ping[3]

[1]College of Pharmacy, Guangdong Pharmaceutical University, Guangzhou, China

[2]School of Physics and Optoelectronic Engineering, Guangdong University of Technology, Guangzhou, China

[3]College of Pharmacy, China Pharmaceutical University, Nanjing, China

ABSTRACT

The objective of the study was to investigate the potential of montmorillonite as a sustained carrier in the preparation of drug-loaded nanoparticles for prolonged ocular application. Nanoparticles were prepared by ionic gelation of chitosan with sodium tripolyphosphate (TPP). Betaxolol hydrochloride (BH) was applied as a model drug. TG, FT-IR, TEM, DLS and XRD have been employed in the characterization of drug-encapsulated chitosan-montmorillonite/TPP nanoparticles (CS-MMT/TPP NPs). TEM images showed that the particles were spherical in shape and had a rough surface. The size range of the nanoparticles was between 338 and 585 nm with positive zeta potential values from 24 mV to 36 mV and encapsulation efficiency values ranging from 12.27% to 50.92%. In vitro sustained drug release was observed with the BH-loaded nanoparticles in artificial tears (pH 7.4). The results of FT-IR, TG and XRD showed that the drug was coated with CS- MMT/TPP NPs. In the mucoadhesion studies, an interaction was found between drug-loaded CS- MMT/TPP NPs and mucin, which could enhance precorneal residence time and hence facilitate an effective sustained release. The optimized formulation was determined to be non-irritant and tolerable by modified Draize test. Therefore, the BH-loading CS-MMT/TPP NPs developed are a promising carrier for controlled drug delivery to the eye.

KEYWORDS

Chitosan, Montmorillonite, Nanoparticles, Mucoadhesive, Sustained Release, Ocular Irritation

1. INTRODUCTION

It is worth mentioning that primary open-angle glaucoma is the most widespread neuropathy, affecting 60.5 million people in the world [1] . Aside from surgery, the only effective treatment for this disease is IOP-reducing drugs applied topically onto the cornea. The necessity of numerous daily applications of eye-drops for life is serious for patients, besides the risk of side effects in the anterior structures of the eye. Betaxolol hydrochloride (BH), and beta-adrenergic blocking agents, are indicated in the treatment of ocular hypertension and chronic open-angle glaucoma [2] . However, there are a number of challenges in the aspect of the treatment of ocular disease due to certain physiological process which may contribute to the poor efficacy of conventional drug formulations. For instance, blinking and tear drainage through the lachrymal drainage system can act to reduce the residence time of topically applied molecule. Eye-drops placed onto the ocular surface are washed in less than 30 seconds [3] . The corneal epithelium is almost impermeable to any substance larger than 500 Da [4] , and most of the commonly used topical drugs are larger than that and consequently cannot cross the cornea. Instead, they permeate throughout the conjunctiva and the underlying sclera in what is known as "non-productive passage". Altogether, less than 5% of topically administered drugs reach intraocular tissues [5] . Therefore, developing and designing a novel delivery platform, which suits the requirement of the topical ocular route is necessary.

Chitosan (CS), a cationic polysaccharide polymer, has some important properties such as mucoadhesivity, biocompatibility, nontoxicity and good ocular tolerance [6] . The mucoadhesive character of the polymer makes it increase the drug residence time due to the capacity to adhere to the mucin coat covering the conjunctiva and the corneal surfaces of the eye by non-covalent bonds forming the basis of ocular mucoadhesion [7] [8] . Moreover, CS solutions show pseudoplastic and viscoelastic properties [9] [10] , which are desirable for ocular drug administration. It has widely being used in ophthalmic preparations [11] . The report by R. S. Bhatta et al. showed that natamycin-loaded chitosan nanoparticles had the ability to prolong the retention time of drug on the ocular surface [12] . The efficiency of the pharmaceutical treatment of most ocular diseases is limited by short ocular residence of conventional pharmaceutical forms of drugs on the ocular mucosa due to the elimination of drug from the corneal surface by the lachrymal flow. To overcome the problem, ocular administration of drugs in the forms of CS nanoparticles has been proposed. CS-based systems can be used for improving the retention and distribution properties of drugs applied topically in the eye. Thus, all preferable properties make it a unique material for the design of ocular delivery vehicles [13] .

Indeed, taking into account the fact that cornea and conjuctiva have a negative charge, it was thought that the use of mucoadhesive polymers which might interact with these extraocular structures would increase the concentration and residence time of the associated drug [14] . The study on cyclosporin A-loaded chitosan nanoparticles has shown the interaction and prolonged residence time of chitosan nanoparticles at the ocular mucosa after the topical administration to rabbits [15] , and gatifloxacin-loaded SA-CH nano-reservoir system has also been investigated for the treatment of ocular infections [14] .

Montmorillonite (MMT) can form many nanocompsites with different organic compounds. Therefore, it expresses a significant capability to be used as a drug carrier for pharmaceutical purpose [16] . The report by Lin et al. has shown that 5-fluorouracil (5-FU) was successfully intercalated into the interlayer of montmorillonite [17] . In addition, the pillared MMT can be intercalated with ibuprofen as a sustained release carrier [18] .

Therefore, MMT material has been proposed as a carrier with high potential for ocular drug delivery.

According to the reports [19] [20] , ionic gelation is the most studied and widely used method for fabricating CS nanoparticles, in which cationic chitosan and multivalent polyanions interact to form CS nanoparticles under simple and mild conditions. Among various polyanions, TPP is the most investigated one due to its quick gelling capacity and non-toxicity property. In the present study, the ionic gelation method was used to prepare BH- loaded CS-MMT/TPP NPs. Different fabricating conditions were evaluated in terms of their effects on physicochemical properties of BH-loaded CS-MMT/TPP NPs, including morphology, particle size, surface charge, the encapsulation efficiency, loading capacity, as well as release profile. MMT and chitosan were chosen as sustained carriers to achieve better encapsulation efficiency and controlled release profiles of optimized nanoparticles. Therefore, the purpose of this study was to develop a suitable and mucoadhesive system consisting of MMT, chitosan and BH for ocular drug delivery.

2. EXPERIMENTAL SECTION

2.1. Materials

Chitosan (deacetylation degree of 95%) was purchased from HAIDEBEI MARINE BIOEEING CO. LTD, Jinan, China. Momtmorillonite was obtained from Zhejiang sanding CO. LTD, china. Betaxolol hydrochloride was purchased from Jinan Haohua Industry, Jinan , China . Male, New Zealand albino rabbits of 2.2 - 2.5 kg were provided by laboratory animal center, Traditional Chinese medicine university of Guangzhou. All other chemicals were of analytical grade.

2.2. Acid Activation of Momtmorillonite

The MMT was activated by treating with 5% H_2SO_4 at the temperature of 40˚C for 1.5 h. The resulting clays were centrifuged and washed with water several times until they were free of SO_4^{2-} , and then were dried at 100 ˚C in an air oven until constant weight was attained.

2.3. Preparation of CS-MMT/TPP NPs

The nanoparticles were prepared by ionic gelation techniques. In the first step MMT was added to acetum (pH = 4.8, 2% v/v) containing betaxolol hydrochloride and heated in water bath at 60°C for 6 hours, and then the solution was ultrasonicated for 5 min and added to chitosan solution and stirred for 1.5 h. subsequently the nanoparticles were fabricated by adding above MMT suspension to chitosan solution (2% v/v) while the TPP solution was added to the mixture suspension with ultrasonication for 10 s and stirring for 60 mins. The dispersed nanoparticles were frozen at −80°C for 48 h, and then freeze-dried at −50°C for 12 h. Finally, the freeze-dried powders were stored in glass vials for further study.

2.4. Entrapment Efficiency (EE) and Drug Loading Capacity (DLC)

To determine the free BH during the preparation of BH-loading CS/TPP NPs and CS-MMT/TPP NPs, the dialyzed solution was collected. The concentration of free BH in the solution was determined by UV-VIS spectrophotometry (UV-1800, MAPADA instrument CO.LTD) using a wavelength of 274.5 nm. The amount of BH in nanoparticles was calculated as the difference between the total amount used to prepare the nanoparticles and the amount present in the supernatant. Fresh prepared BH-loaded CS-MMT/TPP NPs were used for determining the drug-loading capacity. NPs were dissolved in HPLC grade methanol by stirring for 4h, and the solution was ultrasonicated for 5 mins. This colloidal dispersion was filtered by 0.22 μm membrane filter. The drug concentration was analyzed by UV-VIS. The EE and LC of nanoparticles were determined in triplicate and calculated as follows:

$$\text{Encapsulation Efficiency}\,(\text{EE\%}) = \frac{(\text{Total amount of BH-Free BH})}{\text{Total amount BH}} \times 100\% \quad (1)$$

$$\text{Drug loading capacity}\,(\text{DLC\%}) = \frac{(\text{Total amount of BH-Free BH})}{\text{Total weight NPs}} \times 100\% \quad (2)$$

2.5. In Vitro Release Studies

The release of BH from nanoparticles was studied by dialysis method in pH 7.4 artificial tears. 2 ml samples were instilled in the dialysis bag which was screwed with two clamps at each end. The dialysis bag was dipped into the receptor compartment containing 35 ml of dissolution medium and stirred continuously at 100 rpm and maintained at 34°C. The receptor compartment was closed to prevent evaporation of the dissolution medium. Samples were withdrawn at regular time intervals, and the same volume was replaced with fresh dissolution medium. The samples were spectrophotometricially measured at 274.5 nm.

2.6. Particle Size and Zeta Potential

NPs size distribution and zeta potential were determined using photon correlation spectroscopy with a Zetasizer 3000 (DelsaNano S, USA). The size distribution analysis was performed at a scattering angle of 90° at a temperature of 25°C, and zeta potential was measured by electrophoretic light scattering using a disposable zeta cuvette. All measurements were performed in triplicate and the results were presented as mean ± SD.

2.7. Nanoparticles Morphonology

The morphological examination of nanoparticles was performed on transmission electron microscopy (TEM) (JEM-2010HR, Japan). The nanoparticles were applied to a carbon-coated 300 copper grid and then were stained with 2% phosphotungstic acid for viewing by TEM.

2.8. Fourier Transform Infrared Spectroscopy (FT-IR)

FT-IR spectra of freeze-dried nanoparticles were obtained with Bruker Vertex-70 spectrophotometer using potassium bromide (KBr) disk technique. FT-IR spectra of BH-loaded CS-MMT/TPP NPs and their pure components were compared for the evaluation of drug-excipient and excipient-excipient chemical interaction.

2.9. Thermogravimetric Analysis (TG)

TG analysis for BH, CS, MMT, acid-MMT, physical mixture and BH-loaded CS-MMT/TPP NPs were performed using a Netzsch STA 409PC TG instrument with NETZSCH-TA4_5 software Proteus Analysis (Germany). The apparatus was calibrated with purified indium (99%). The samples were placed in flat-bottomed aluminum pan and heated at a constant rate of 10°C/min in nitrogen atmosphere in a temperature range of 30°C - 1000°C.

2.10. X-Ray Diffratometry

X-ray diffraction patterns for BH, CS, acid-MMT, physical mixture, BH-loaded CS-MMT/TPP NPs were obtained using a Bruker D8 Advance X-ray diffractometer equipped with CuKα. The tube was operated at 40 KV and 40 mA.

2.11. Mucoadhesion Studies

Equal volumes of mucin solution and nanoparticles were vortexed for 1 min and zeta potential of the mixture was measured by photon correlation Spectroscopy with a Zetasizer 3000 (DelsaNano S, USA). In this study, the mucoadhesive properties of nanoparticles were evaluated by measuring the change of zeta potential on interaction with negatively charge mucin [21]. The alteration of zeta potential of nanoparticles indicates the interaction with mucin.

2.12. Ocular Irritation Studies

The ocular irritancy and damaging effects of NPs and BH were evaluated according to a modified Draize test [22] . Test substance (50 ul) was instilled directly to the cornea in the right eye every 12 h for 7 days (14 treatments), while left eye severed as control and was treated with normal saline. Observation was carried out before the treatment for every day and was done to evaluate ocular tissue after the last treatment of 1, 2, 4, 24, 48 and 72 h. The congestion, swelling, discharge and redness of the conjunctiva were grade on a scale from 0 to 3, 0 to 4, 0 to 3, and 0 to 3, respectively. Irritation and cornea opacity were grade on a scale from 0 to 4.

2.13. Statistical Analysis

The average difference of data was compared by one-way analysis of variance or independent sample t-test (origin 8.5 USA). All data are presented as a mean value with its standard deviation indicated (mean ± SD). P-values less than 0.05 were considered to be statistical significance.

3. RESULT AND DISCUSSION

3.1. Entrapment Efficiency (EE) and Drug Loading Capacity (DLC)

The effect of CS concentration on the encapsulation efficiency and drug-loading capacity were examined as shown in Table 1. The increasing concentration of CS from 1.0 mg/ml to 1.5 mg/ml at a constant TPP concentration and chitosan to TPP mass ratio, caused a great increase in drug content from 30.32% to 39.51%. This indicates that a less amount of nanoparticles were formed due to the lower chitosan concentration. However, when the concentration of CS was increased from 1.5 mg/ml to 3.0 mg/ml, a significant decrease in the encapsulation efficiency from 39.51% to 31.6% was exhibited. This may be explained by the fact that relatively higher viscosity of the higher CS concentration hinders the drug encapsulation [23] . With the similar results, the loading capacity firstly increased with the increase of the CS concentration and then decreased. The reason appears to be similar to the above definition. In addition, the increase of the weight of CS and the corresponding change of encapsulation efficiency were a contributor to the result as well. The effects of chitosan to TPP mass ratio on the drug content and loading capacity of nanoparticles were studied at the mass ratio of 10:1, 10:1.5, 10:2, 10:2.5 and 10:3 at a fixed TPP concentration of 1.0 mg/ml, and the results were presented in Table 2. The encapsulation efficiency and loading capacity of nanoparticles decreased with the decrease in weight ratio of CS/TPP from 10:1 to 10:2. A possible explanation may be that a more compact solid-matrix structure due to the increased mass ratio for CS/TPP, which leads to the increasing amount of formed nanoparticles, resulting in the increase for encapsulation efficiency and loading capacity in nanoparticles. However, when the mass ratio is over the critical point (10:2), the aggregation of nanoparticles occurs, which indicated the reduction of encapsulation efficiency. The influence of the different amount of MMT on drug content and drug loading capacity was also examined, as shown in Table 3. Table 3 shows that encapsulation efficiency increased from 5.5% to 9.21%, when initial weight of MMT increased from 1 mg to 5 mg. However, the further increase of MMT amount does not lead to a sequential increase of encapsulation efficiency. The reason may be the more drugs adsorbed and intercalated with the higher MMT weight could be a contributor to the increase. When the amount of MMT was increased to some value, the drug content appears to keep a constant value possibly due to the redundant MMT which does not participate in the formation of the nanoparticles. The effect of initial BH concentration on drug content and loading capacity of the nanoparticles were also examined, as shown in Table 4. The increase of BH concentration was done from 0.75 mg/ml to 1.25 mg/ml, accompanied with the increasing encapsulation from 3.12% to 12.44% (Table 4). However, when BH concentration reaches 1.25 mg/ml, the drug content tends to decrease with progressive enhancement of drug loading capacity. This indicates that BH adsorption and encapsulation by electrostatic interactions normally exhibit a saturation kinetics which reaches a peak value at high concentration under constant temperature. On the basis of single-factor test, orthogonal test was carried out by choosing four factors and each factor had three levels (Table 5). Preparation technology was optimized through the orthogonal experiment. According to the analysis of variance, the influence factors of drug content of the nanoparticles were followed by the sequence: TPP concentration > chitosan concentration > drug concentration > chitosan to TPP mass ratio. The optimized preparation was chosen for further study.

3.2. Effect of Chitosan Concentration on In-Vitro Release Performance

Figure 1 shows BH release profiles in the respect of different chitosan concentration during BH encapsulation. The reduction of initial burst release with the chitosan concentration increasing from 1.0 mg/ml to 2.0 mg/ml during BH encapsulation was observed. This may be explained by the fact that the increased viscosity at higher chitosan concentration allowed the formation of denser nanoparticles upon the interaction with TPP, resulting in greater crosslink density and less swelling ability, and consequently reducing the BH release [24] . In addition, the decrease of encapsulation efficiency with the increase of chitosan concentration may be a contributor to this phenomenon. With the durative increase chitosan concentration, the increase of initial burst release for nanoparticles was observed due to the aggregation of nanoparticles caused by the increased chitosan concentration resulting in the more loose structure for nanoparticles.

3.3. Effect of Amount Varies of Montmorillonite on In-Vitro Release Performance

The effect of different amount of montmorillonite on BH release from prepared nanoparticles was studied. From the release profiles in Figure 2, we can see that the overall rate of drug release (i.e. the release fraction of drug content of nanoparticles at a given time) tended to become lower as the amount of MMT increasing from 1 mg to 5 mg, while it started to become higher with the further increase of the weight of MMT. It is obvious that the amount of MMT should be in an appropriate range, which can only effectively form nanoparticles with chitosan and TPP. This indicates that MMT at high amount would interrupter the formation of denser chitosan nanoparticles due to the hydroxyl group in the MMT interacting with the amino of chitosan, which disturbed the electrostatic interaction between chitosan and TPP. These reasons resulted in more fast release of BH from nanoparticles.

The result of drug release experiment for BH solution and optimized NPs in stimulated ocular circumstances (37°C, pH 7.4) are shown in Figure 3. We can observe that BH-loaded CS-MMT/TPP NPs showed a lower burst release as compared to BH-loaded CS/TPP NPs, but both presented a complete release of BH over a period of 10 h. For pure BH solution, which showed a fast release profile with 100% BH released within 2 h. Hence, NPs exhibited significantly sustained release in comparison to drug solution. In addition, this could also indicate that CS-MMT/TPP NPs showed better sustained-release compared to that of CS/TPP NPs. From the release profiles of NPs, we may come to a conclusion that BH encapsulated in the matrices of nanoparticles are released by three mechanisms taking place in sequence: 1) BH desorption from the particle surface; 2) diffusion and readsorption of the BH through the pores of the polymer network or the interlayer of montmorillonite; 3) degradation and erosion of the polymeric network.

Table 1. Effect of chitosan concentration on the encapsulation efficiency and drug loading capacity for the prepared nanoparticles (mean ± SD).

Formula code	CS concentration (mg/ml)	Encapsulation efficiency (%)	Loading capacity (%)
1	1	30.32 ± 0.12	11.49 ± 0.33
2	1.5	39.51 ± 0.23	11.99 ± 0.26
3	2	33.32 ± 0.13	8.67 ± 0.45
4	2.5	33.01 ± 0.21	6.68 ± 0.78
5	3	31.16 ± 0.11	5.99 ± 0.13

Particles preparation conditions: the weight of montmorillonite = 5 mg, BH concentration = 1.25 mg/ml, chitosan/TPP = 10:3, pH = 4.8.

Table 2. Effect of chitosan to TPP mass ratio on the encapsulation efficiency and drug loading capacity for the prepared nanoparticles (mean ± SD).

Formula code	Mass ratio of CS to TPP	Encapsulation efficiency (%)	Loading capacity (%)
1	10:1	45.63 ± 0.11	13.87 ± 0.78
2	20:3	43.77 ± 0.36	12.64 ± 0.34
3	10:2	39..82 ± 0.23	11.58 ± 0.46
4	20:5	39.41 ± 0.21	11.49 ± 1.26
5	10:3	36.51 ± 0.15	11.15 ± 0.27

Particles preparation conditions: the weight of montmorillonite = 5 mg, BH concentration = 1.25 mg/ml, chitosan concentration = 2 mg/ml, pH = 4.8.

Table 3. Effect of the amount of MMT on the encapsulation efficiency and drug loading capacity for the prepared nanoparticles (mean ± SD).

Formula code	MMT (mg)	Encapsulation efficiency (%)	Loading capacity (%)
1	1	19.29 ± 0.25	5.5 ± 0.21
2	3	26.96 ± 0.32	7.55 ± 0.34
3	5	35.47 ± 0.12	9.21 ± 0.46
4	7	37.13 ± 0.45	9.38 ± 0.67
5	10	37.75 ± 0.79	9.48 ± 0.43

Particles preparation condition: chitosan concentration = 2 mg/ml, BH concentration = 1.25 mg/ml, chitosan/TPP = 10:3, pH = 4.8.

Table 4. Effect of various initial drug concentration on the encapsulation efficiency and drug loading capacity for prepared nanoparticles (mean ± SD).

Formula code	Drug concentration (mg/ml)	Encapsulation efficiency (%)	Loading capacity (%)
1	0.75	11.92% ± 0.13%	3.12% ± 0.12%
2	1.0	17.20% ± 0.21%	5.38% ± 0.25%
3	1.25	34.72% ± 0.11%	12.44% ± 0.01%
4	1.5	23.79% ± 0.26%	10.20% ± 0.77%
5	2.0	21.65% ± 0.33%	10.09% ± 0.39%

Particles preparation conditions: the weight of montmorillonite = 5 mg, chitosan concentration = 1.25 mg/ml, chitosan: TPP = 10:3, pH = 4.8.

Table 5. Effect of chitosan concentration, TPP concentration, mass ratio of chitosan/TPP and BH concentration on the encapsulation efficiency, loading capacity, particle size, zeta potential values and PID (mean ± SD).

code	A[a]	B[b]	C[c]	D[d]	EE (%)	LC (%)	Particle size (nm)	Zeta Potential (mV)	PID
1	1.0	0.8	10:1	0.75	13.10 ± 0.21	5.26 ± 0.16	338.4 ± 2.3	25.14 ± 3.6	0.254 ± 0.11
2	1.0	1.0	10:1.5	1.25	15.17 ± 0.11	8.20 ± 0.34	366.5 ± 5.6	30.31 ± 2.7	0.213 ± 0.08
3	1.0	1.2	10:2.5	2	18.00 ± 0.32	11.49 ± 0.31	585.5 ± 7.8	24.62 ± 1.2	0.282 ± 0.23
4	1.5	0.8	10:1.5	2	19.53 ± 0.23	10.56 ± 0.24	355.8 ± 4.3	29.49 ± 4.8	0.222 ± 0.34
5	1.5	1.0	10:2.5	0.75	12.27 ± 0.13	3.49 ± 0.38	529.3 ± 2.2	31.82 ± 2.1	0.260 ± 0.21
6	1.5	1.2	10:1	1.25	35.31 ± 0.42	14.85 ± 0.56	480.5 ± 4.5	30.70 ± 3.9	0.282 ± 0.13
7	2.0	0.8	10:2.5	1.25	36.11 ± 0.35	11.28 ± 0.43	503.5 ± 5.9	25.27 ± 4.2	0.242 ± 0.22
8	2.0	1.0	10:1	2	13.40 ± 0.12	6.33 ± 0.57	451.7 ± 6.7	29.40 ± 5.6	0.220 ± 0.26
9	2.0	1.2	10:1.5	0.75	50.92 ± 0.48	11.59 ± 0.24	524.3 ± 8.2	36.42 ± 2.3	0.275 ± 0.15

[a]Chitosan concentration; [b]TPP concentration; [c]Chitosan/TPP mass ratio; [d]BH concentration.

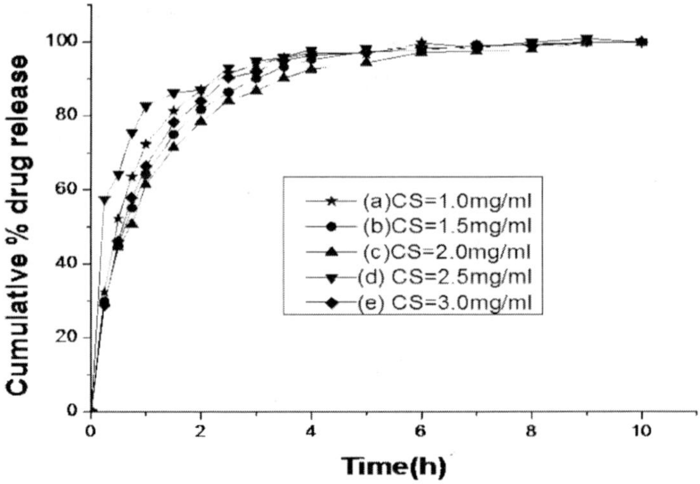

Figure 1. The influence of different chitosan concentration (a) chitosan = 1.0 mg/ml; (b) chitosan = 1.5 mg/ml; (c) chitosan = 2.0 mg/ml; (d) chitosan = 2.5 mg/ml; (e) chitosan = 3 mg/ml) on BH release from prepared nanoparticles.

3.4. Particle Size and Zeta Potential

The reduced size of nanoparticle formulation was very interesting in the view of their application in ocular drop formulations. It is well known that the size of particle under 10 μm would not cause any ocular irritation. Therefore, the reduction achieved in particles size could result in more compliance for patient. In this study, all the nanosystems ranged in diameter from 338.4 ± 2.3 to 585.5 ± 7.8 nm with a polydispersity index values between 0.213 ± 0.08 to 0.282 ± 0.13, as shown in Table 5.

The effect of concentration and weight ratio of CS and TPP on the mean nanoparticle diameter were examined. The nanoparticle diameter appears to be dependent on the concentration of CS and TPP and chitosan to TPP mass ratio, the minimum diameter (338.4 nm) corresponding to the lowest CS and TPP concentration, and maximum diameter (585.5 nm) corresponding to the highest TPP concentration and weight ratio between chitosan and TPP. The increase CS concentration from 1.0 mg/ml for formula 1 to 1.5 mg/ml for formula 4, then 2.0 mg/ml for formula 7 at the constant TPP concentration, showed a significant increase in the particle diameter from 338 to 355, then 503 nm, respectively. The increased particle diameters were attributed to the increase of CS viscosity [25] and the higher availability of protonated amine groups for ionic gelation with increasing CS concentration [14] .

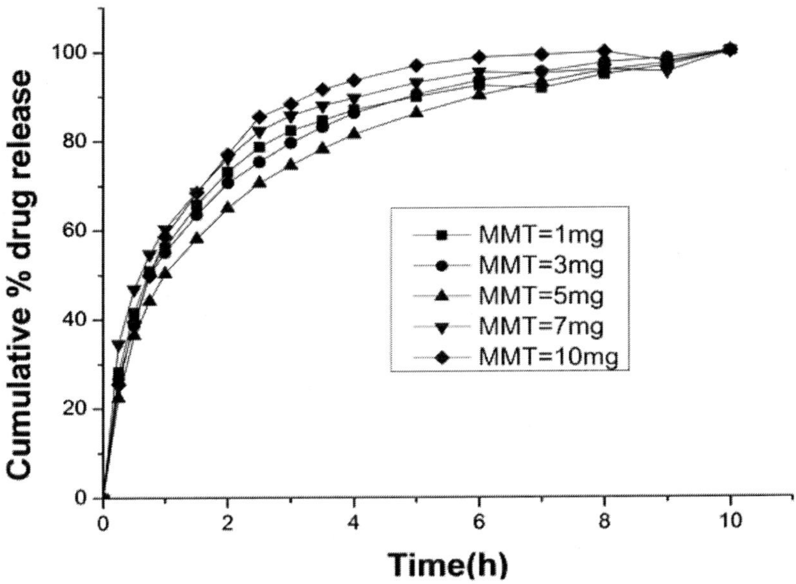

Figure 2. In vitro release profiles of different amount of acid MMT prepared nanoparticles.

Zeta potential studies of the nanoparticles showed the values ranging from 24.62 ± 1.2 to 36.42 ± 2.3 mv (Table 5). The positive zeta-potential values indicate that the surface of the nanosystems is mostly composed of CS [26] . This net positive charge of the particle is desirable to promote electrostatic interaction with the ocular mucin [27] . In general, it was noted that zeta potential of the nanoparticles decreased as the CS/TPP ratio decreased. The lower zeta potential with increasing TPP amounts might be caused by an increased masking of free positively changed amino groups of CS. In addition, increasing CS concentration results in an increase of zeta potential of the nanoparticles due to the increase of free positively charged amino groups of CS. According to the analysis of variance (data not shown), the effect factors on nanoparticles size and zeta potential take place in sequence: chitosan to TPP mass ratio > TPP concentration > chitosan concentration > drug concentration and chitosan to TPP mass ratio > chitosan concentration > TPP concentration > drug concentration, respectively.

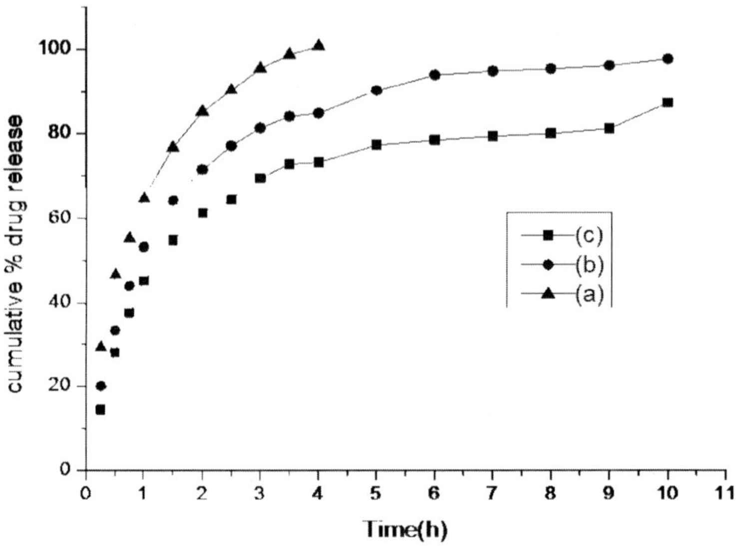

Figure 3. In vitro release profiles of (a) BH, (b) BH loading CS/TPP NPs,(c) BH loading CS-MMT/TPP NPs.

3.5. Nanoparticles Morphology

A representative TEM image for drug-loaded CS-MMT/TPP nanoparticles was presented in Figure 4. The results indicated that the tested nanoparticles were predominantly spherical in the shape with an irregular surface. The diameter of nanoparticles was about 250 nm smaller than the diameter determined by photon correlation spectroscopy, in which this could have been expected since the fresh prepared nanoparticles suspensions for the photon correlation spectroscopy experiments, and the chitosan and MMT have the ability to swell in contact with acetic acid, while the TEM experiments were performed with dry samples.

3.6. Fourier Transforms Infrared Spectroscopy (FT-IR)

BH, chitosan, montmorillonite, their physical mixture, BH-loaded CS/TPP NPs and BH-loaded CS-MMT/TPP NPs were analyzed using FT-IR spectrophotometer for characteristic absorption bands (Figure 5). The IR spectrum of BH showed several obvious characteristic peaks: the strong peak at 3248 cm^{-1} belongs to the -OH stretching vibrations; aromatic C=C stretching vibrations are at 1610 cm^{-1} and 1520 cm^{-1}; aromatic C-H stretching vibrations are at 3089 cm^{-1}; the peaks at 2968 cm^{-1} and 2961 cm^{-1} correspond to $-CH_3$ stretching; the peaks at 2876 cm^{-1} and 2812 cm^{-1} correspond to $-CH_2$stretching; C-O-C stretching vibrations are at 1084 cm^{-1}. In MMT spectrum the strong peak at 3420 cm^{-1} corresponds to the O-H stretching vibrations; Si-O stretching is at 1102 cm^{-1} and 1042 cm^{-1}; the intense peaks of 519 cm^{-1} and 479 cm^{-1} belong to Fe-O-Si stretching and Mg- O-Si stretching, respectively; the stretching of Mg-Al-OH is at 782 cm^{-1}; The IR spectrum of chitosan was characterized by the intense peak at 3435 cm^{-1} corresponded to the combined peak of O-H stretching and intermolecular hydrogen bonding. Specific peaks of chitosan were observed at 1630 cm^{-1} (amide I) and 1593 cm^{-1} (amine II). The symmetric stretch of C-O-C is found at around 1077 cm^{-1}, and the intense peak at 1416 cm^{-1} belongs to C-N stretching.

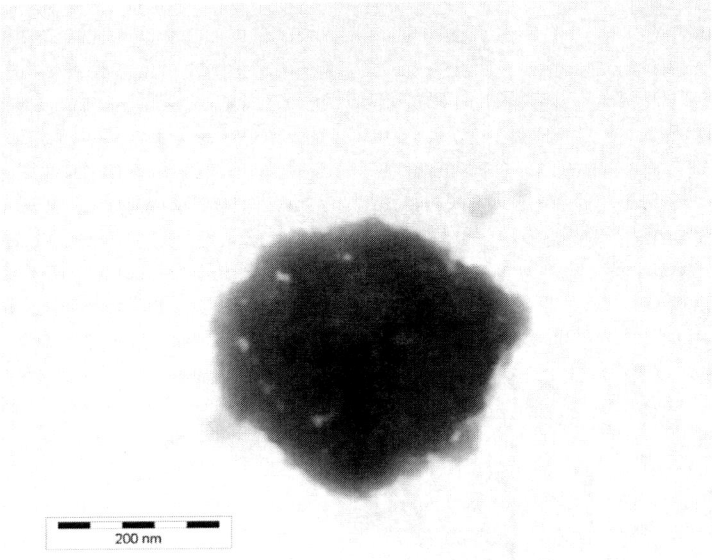

Figure 4. Transmission electron micrograph of BH-loaded CS-MMT/TPP NPs, bar represent 1 µm.

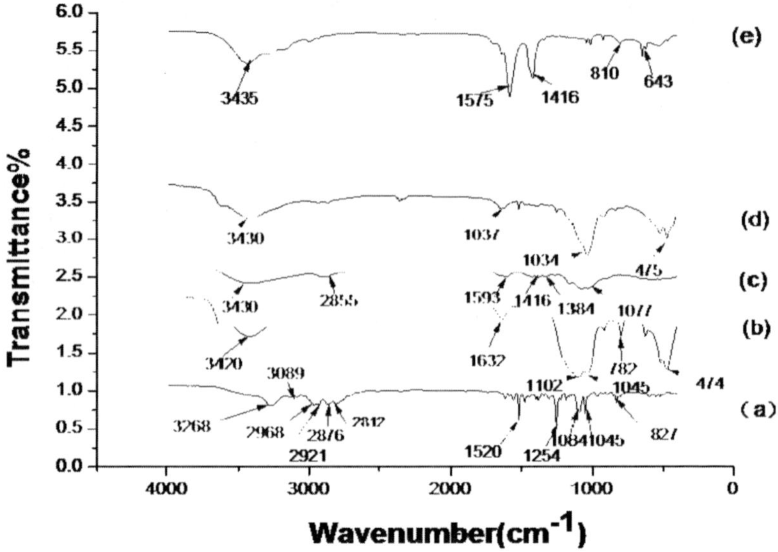

Figure 5. FTIR spectrum of (a) BH, (b) MMT, (c) CS, (d) Physical mixtures of BH, MMT and CS, (e) BH-loaded CS-MMT/TPP NPs.

The spectrum patterns of physical mixtures corresponded simply to the superposition of the IR spectra of betaxolol, montmorillonite and chitosan. The characteristic peaks for BH were almost completely obscured by the very intense and broad peaks of chitosan and montmorillonite. The FT-IR spectrum of BH-loaded CS-MMT/ TPP NPs is different from that of physical mixtures and each pure component. In NPs the tip of peak at 3420 cm^{-1} was shifted to 3435 cm^{-1}, indicating that hydrogen bonding is enhanced. The 1953 cm^{-1} peak of N-H blending vibration disappeared and new peaks at 1575 cm^{-1} and 1416 cm^{-1} were observed. This spectra changed may be attributed to the electrostatic interaction between chitosan and TPP. In addition, the hydrogen-bond interaction between chitosan and montmorillonite could explain this fact as well.

3.7. Thermogravimetric Analysis (TG)

The TG curves of pure components, their corresponding physical mixture, BH-loaded CS/TPP NPs and BH- loaded CS-MMT/TPP NPs were showed in

Figure 6. There are two distinct changes in the TG profiles of montmorillonite. The first weight loss from 42°C to 184°C may be due to the free water evaporation. The second weight loss from 480°C to 625°C is attributed to the structural dehydration. In the curves of simple physical mixture, the weight loss from 228°C to 347°C corresponds to the decomposition of BH. Other weight losses are the same as those of MMT and CS. The TG patterns of MMT-BH/CS NPs are shown in Figure 6(d). The first weight loss from 42°C to 184°C may be attributed to the free water evaporation. The second weight loss from 228°C to 347°C may be owed to free drug decomposition. The third weight loss from 450°C to 520°C is attributed to the drug decomposition which was encapsulated into nanoparticles. We can see that the weight loss at around 300°C in Figure 6(e) shifts to 500°C in Figure 6(d), which may be explained by the fact that the BH is partly encapsulated into the nanoparticles. The last weight loss from 840°C to 900°C may be due to the destruction of ionic bond between CS and TPP.

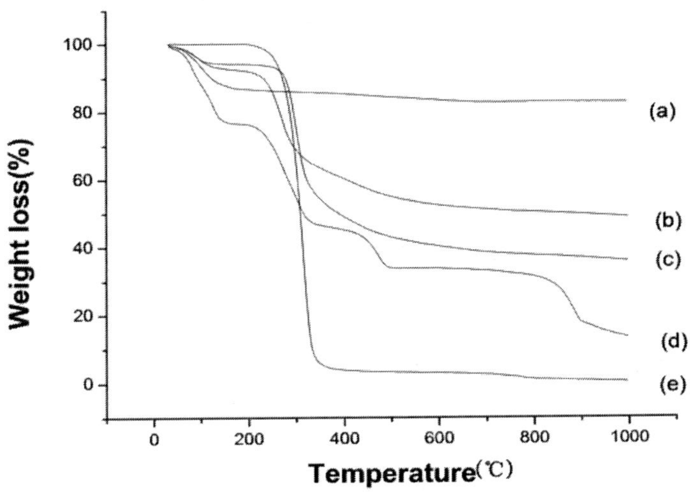

Figure 6. TGA curves of (a) MMT, (b) physical mixture of BH, MMT, CS, (c) CS, (d) BH loading CS-MMT/TPP NPs, (e) BH.

3.8. X-Ray Diffratometry

Figure 7 shows XRD patterns of BH, acid-MMT, CS, their physical mixture, blank nanoparticles and optimized nanoparticles. BH has a specific sharp peak. MMT and CS show the topical reflection at 2-theta = 5.82 and 20.18, respectively. No characteristic peaks of CS and MMT were found in diffracograms of blank nanoparticles and optimized nanoparticles, which indicated that chitosan nanoparticles were comprised of a dense network structure of interpenetrating polymer chain crosslinked to each other by the electrostatic reaction between TPP and chitosan. The XRD demonstrated the greater disarray in chain alignment in the nanoparticles after cross-link. Therefore, the BH-loaded CS-MMT/TPP NPs were successfully formed.

3.9. Mucoadhesion Studies

Mucoadhesion studies of the optimized nanoparticles showed that the nanoparticles revealed the reduction in zeta potential values for CS nanoparticles occurring on mixing with mucin (Figure 8). The reduction may be attributed to the ionic reaction between negatively charged sialic acid residues in mucin and positively charged amino groups in CS nanoparticles that results in the mucoadhesive properties of the tested nanoparticles. BH/CS nanoparticles showed more reduction in zeta potential values indicating more mucoadhesive compared to BH- loaded CS-MMT/TPP NPs. This may be attributed to the fact that BH-loaded CS/TPP NPs had higher zeta potential values (+34 mV) than MMT-BH/CS nanoparticles (+29.18 mV).

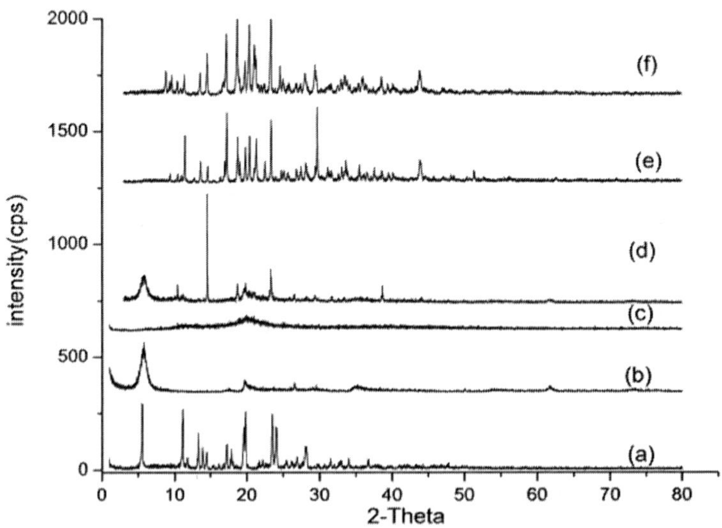

Figure 7. X-ray difffraction of (a) BH; (b) acid-MMT; (c) CS; (d) their physical mixture; (e) blank NPs and (f) oputimized NPs.

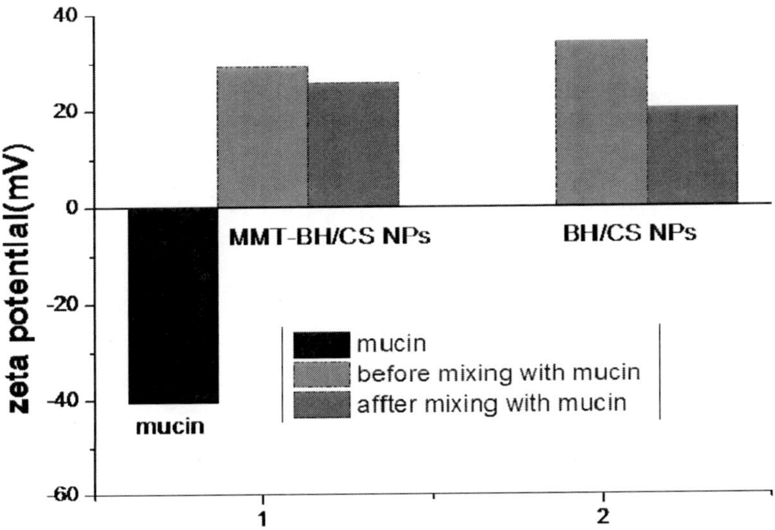

Figure 8. Zeta potential values obtained for mucin and BH/CS and MMT-BH/CS nanoparticles before and after incubation with mucin.

3.10. Ocular Irritation Studies

Figure 9 presents the histopathology of the tested rabbit's eyeballs with various preparations to investigate their influence on cell structure and tissue integrity. The in vivo study showed no sign of irritation or damaging effects in cornea, conunctiva or iris. The scores for conjunctiva congestion of MMT-CS/TPP nanoparticle formulation and control group were grade zero. The cornea and iris scores were zero for all tested samples at all observations (data not shown). Pathology of rabbit eyeballs and lids confirmed the presence of normal ocular surface structure in both optimized nanoparticles cornea and conjunctival epithelial cell maintained normal morphology and integrated epithelium. The basal columnar cell of cornea can be recognized as single lager cell, and is normally packed by tight junction complex (Figure 9(a) & Figure 9(b)). Similarly, normal levels of scattered polymorphonuclear cells were observed in the conjuctival stroma (Figure 9(c) & Figure 9(d)), which indicated that there were no signs of inflammation and tissue edema. The treatment of corneas with optimized formulations was exemplified in Figure 9(a), and the histopathology confirmed that no ocular irritating effects were induced by MMT-CS/TPP nanoparticle formulation compared with control group. All tested samples showed a good corneal biocompatibility which demonstrated a preferable ocular tolerance.

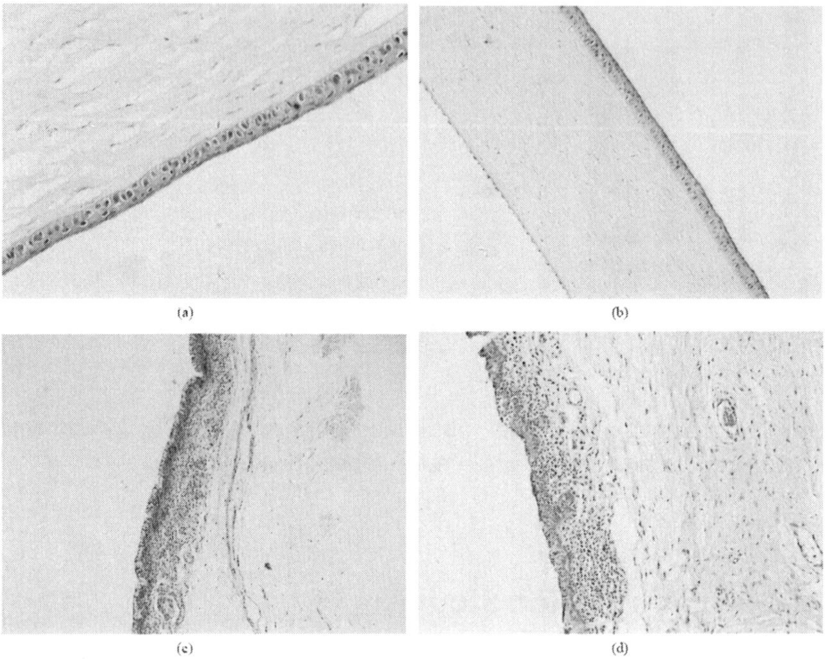

Figure 9. Histopathology microscopy of the ocular tissues including cornea (a) and (b) and conjunctiva (c) and (d) after treated with different solutions for 7 days. (a) and (c) MMT-BH/CS NPs; (b) and (d) Normal saline.

4. CONCLUSION

In the present study, the BH loading CS-MMT/TPP nanoparticles were prepared successfully by ionic reaction between CS and TPP based on the positively charged amino groups in CS and the negatively charged phosphate groups in TPP. In addition, the hydroxide radical on the surface of montmorillonite interacting with chitosan through hydrogen bond also played an important role during the preparation of nanoparticles. The drug-loaded nanoparticles consisting of MMT, CS and TPP, showed the enhancement of precorneal retention, sustained release, and high drug-loading capacity. The ratio between TPP and CS was critical for the control of particle size and encapsulation efficiency of nanoparticles. No ocular damage in cornea, conjunctiva or iris was observed. Consequently, these nanocarriers represented a promising approach for the circumvention of the present limitation in ocular drug delivery.

ACKNOWLEDGEMENTS

The authors would thank to the youth Natural Science Foundation for financial support (NO.51102052), and thank the eye institute of Shandong province in China for the provision of human immortalized cornea epithelial cell line (HCEC).

REFERENCES

1. Quigley, H.A. and Broman, A.T. (2006) The Number of People with Glaucoma Worldwide in 2010 and 2020. British Journal of Ophthalmology, 90, 262-267. http://www.ncbi.nlm.nih.gov/ pmc/ articles/PMC1856963/

2. Mahalaxmi, D., Samarth, M.M., Shiravadekar, H.S. and Sanghavi, N.M. (1996) Stability-Indicating HPLC Method for Betaxolol HCl and Its Pharmaceutical Dosage Forms. Drug Development and Industrial Pharmacy, 22, 1037-1039.

3. Kaur, I.P. and Kanwar, M. (2002) Ocular Preparations: The Formulation Approach. Drug Development and Industrial Pharmacy, 28, 473-493.

4. Hamalainen, K.M., Kananen, K., Auriola, S., Kontturi, K. and Urtti, A. (1997) Characterization of Paracellular and Aqueous Penetration Routes in Cornea, Conjunctiva and Sclera. Investigative Ophthalmology and Visual Science, 38, 627-634.

5. Keister, J.C., Cooper, E.R., Missel, P.J., Lang, J.C. and Hager, D.F. (1991) Limits on Optimizing Ocular Drug Delivery. Journal of Pharmaceutical Sciences, 80, 50-53.

6. Calvo, C.P., Vila-Jato, J.L. and Alonso, M.J. (1997) Novel Hydrophilic Chitosan-Polyethylene Oxide Nanoparticles as Protein Carriers. Journal of Applied Polymer Science, 63, 125-132.

7. Kaur, I.P. and Smitha, R. (2002) Penetration Enhancers and Ocular Bioadhesives: Two New Avenues for Ophthalmic Drug Delivery. Drug Development and Industrial Pharmacy, 28, 353-369. http://informahealthcare.com/doi/abs/10.1081/DDC-120002997%20

8. Ludwig, A. (2005) The Use of Mucoadhesive Polymers in Ocular Drug Delivery. Advances Drug Delivery Reviews, 57, 1595-1639. http://www.ncbi.nlm.nih.gov/pubmed/16198021

9. Mucha, M. (1997) Rheological Characteristics of Semi-Dilute Chitosan Solutions. Macromolecular Chemistry and Physics, 198, 471-484. http://onlinelibrary.wiley.com/doi/10.1002/macp.1997.021980220/abstract

10. Wang, W. and Xu, D. (1994) Viscosity and Flow Properties of Concentrated Solutions of Chitosan with Different Degrees of Deacetylation. International Journal of Biological Macromolecule, 16, 149-152. http://www.ncbi.nlm.nih.gov/pubmed/?term

11. Alonso, M.J. and Sanchez, A. (2003) The Potential of Chitosan in Ocular Drug Delivery. Journal of Pharmacy and Pharmacology, 55, 1451-1463. http://onlinelibrary.wiley.com/doi/10.1211/0022357022476/abstract;jsessionid=8714D64E9E7D3894BF48FBA2B0D0912C.f01t04

12. Bhatta, R.S., Chandasana, H., Chhonker, Y.S., Rathi, C., Kumar, D., Mitra, K. and Shukla, P.K. (2012) Mucoadhesive Nanoparticles for Prolonged Ocular Delivery of Natamycin: In Vitro and Pharmacokinetics Studies. International Journal of Pharmaceutics, 432, 105-112. http://www.sciencedirect.com/science/article/pii/S0378517312004061

13. Diebold, A. and Calonge, M. (2010) Applications of Nanoparticles in Ophthalmology. Progress in Retinal and Eye Research, 29, 596-609. http://www.sciencedirect.com/science?_ob=ArticleListURL&_method=list&_ArticleListID=-782671355&_sort=r&_st=13&view=c&md5=e5bfcd6370882023e9fc1c79a37ae974&searchtype=a

14. Motwani, S.K., Chopra, S., Talegaonkar, S., Kohli, K., Ahmad, F.J. and Khar, R.K. (2008) Chitosan-Sodium Alginate Nanoparticles as Submicroscopic Reservoirs for Ocular Delivery: Formulation,

Optimization and in Vitro Characterization. European Journal of Pharmaceutics and Biopharmaceutics, 68, 513-525. http://linkinghub.elsevier.com/retrieve/pii/S0939-6411(07)00308-6

15. Angela, M., Campos, D., Sanchez, A. and Alonso, M.J. (2001) Chitosan Nanoparticles: A New Vehicle for the Improvement of the Delivery of Drugs to the Ocular Surface. Application to Cyclosporin A. International Journal of Pharmaceutics, 224, 159-168. http://linkinghub. elsevier. com/retrieve/pii/S0378517301007608

16. Ennajih, H., Bouhfid, R., Essassi, E.L.M., Bousmina, M. and El Kadib, A. (2012) Chitosan-Montmorillonite Bio-Based Aerogel Hybrid Microspheres. Microporous and Mesoporous Materials, 152, 208-213.

17. Lin, F.H., Lee, Y.H., Jian, C.H., Wong, J.M., Schiech, M.J. and Wang, C.Y. (2002) A Study of Purified Montmorillonite Intercalated with 5-Fluorouracil as Drug Carrier. Biomaterials, 23, 1981-198. http://www.ncbi.nlm.nih.gov/pubmed/11996039

18. Zheng, J.P., Luan, L., Wang, H.Y., Xi, L.F. and Yao, K.D. (2007) Study on Ibuprofen/Montmorillonite Intercalation Composites as Drug Release System. Applied Clay Science, 36, 297-301. http://www.sciencedirect. com/science/article/pii/S0169131707000026

19. Bhatta, R.S., Chandasana, H., Chhonker, Y.S., Rathi, C., Kumar, D., Mitra, K. and Shukla, P.K. (2012) Mucoadhesive Nanoparticles for Prolonged Ocular Delivery of Natamycin: In Vitro and Pharmacokinetics Studies. International Journal of Pharmaceutics, 432, 105-112. http://www. sciencedirect.com/science/article/pii/S0378517312004061

20. Qi, L.F., Xu, Z.R., Jiang, X., Hu, C.H. and Zou, X.F. (2004) Preparation and Antibacterial Activity of Chitosan Nanoparticles. Carbohydrate Research, 339, 2693-2700. http://www.ncbi.nlm.nih.gov/pubmed/15519328

21. Mahmoud, A.A., El-Feky, G.S., Kamel, R. and Awad, G.E.A. (2011) Chitosan/Sulfobutylether-β-Cyclodextrin Nanoparticles as a Potential Approach for Ocular Drug Delivery. International Journal of Pharmaceutics, 413, 229-236. http://linkinghub.elsevier. com/retrieve/pii/S0378-5173(11)00358-9

22. Wilhelmus, K.R. (2001) The Draize Eye Test. Survey of Ophthalmology, 45, 493-515. http://www.ncbi.nlm.nih.gov/pubmed/11425356

23. Vandenberg, G.W., Drolet, C., Scott, S.L. and Noue, J.D. (2001) Factors Affecting Protein Release from Alginate- Chitosan Coacervate Microcapsules during Production and Gastric/Intestinal Simulation. Journal of Controlled Release, 77, 297-307. http://www.sciencedirect.com/science/article/pii/S016836590100517X

24. Quan, G. and Tao, W. (2007) Chitosan Nanoparticle as Protein Delivery Carrier—Systematic Examination of Fabrication Conditions for Efficient Loading and Release. Colloid and Surfaces B: Biointerfaces, 59, 24-34. http://www.sciencedirect.com/science/article/pii/S0927776507001671

25. Kawashima, Y., Yamamoto, H., Takeuchi, H. and Kuno, Y. (2000) Mucoadhesive DL-Lactide/Glycolide Copolymer Nanoparticles Coated with Chitosan to Improve Oral Delivery of Elcatonin. Pharmaceutical Development and Technology, 5, 77-85. http://informahealthcare.com/doi/abs/10.1081/PDT-100100522%20

26. Oyarzun-Ampuero, F.A., Brea, J., Loza, M.I., Torres, D. and Alonso, M.J. (2009) Chitosan-Hyaluronic Acid Nanoparticles Loaded with Heparin for the Treatment of Asthma. International Journal of Pharmaceutics, 2, 122-129. http://www.sciencedirect.com/science/article/pii/S037851 7309002129

27. Schiffelers, R.M., Woodle, M.C. and Scaria, P. (2004) Pharmaceutical Prospects for RNA Interference. Pharmacy Research, 21, 1-7. http://link.springer.com/article/10.1023%2FB%3APHAM.0000012145.4 9054.6c

Index